S A C R A M E N T

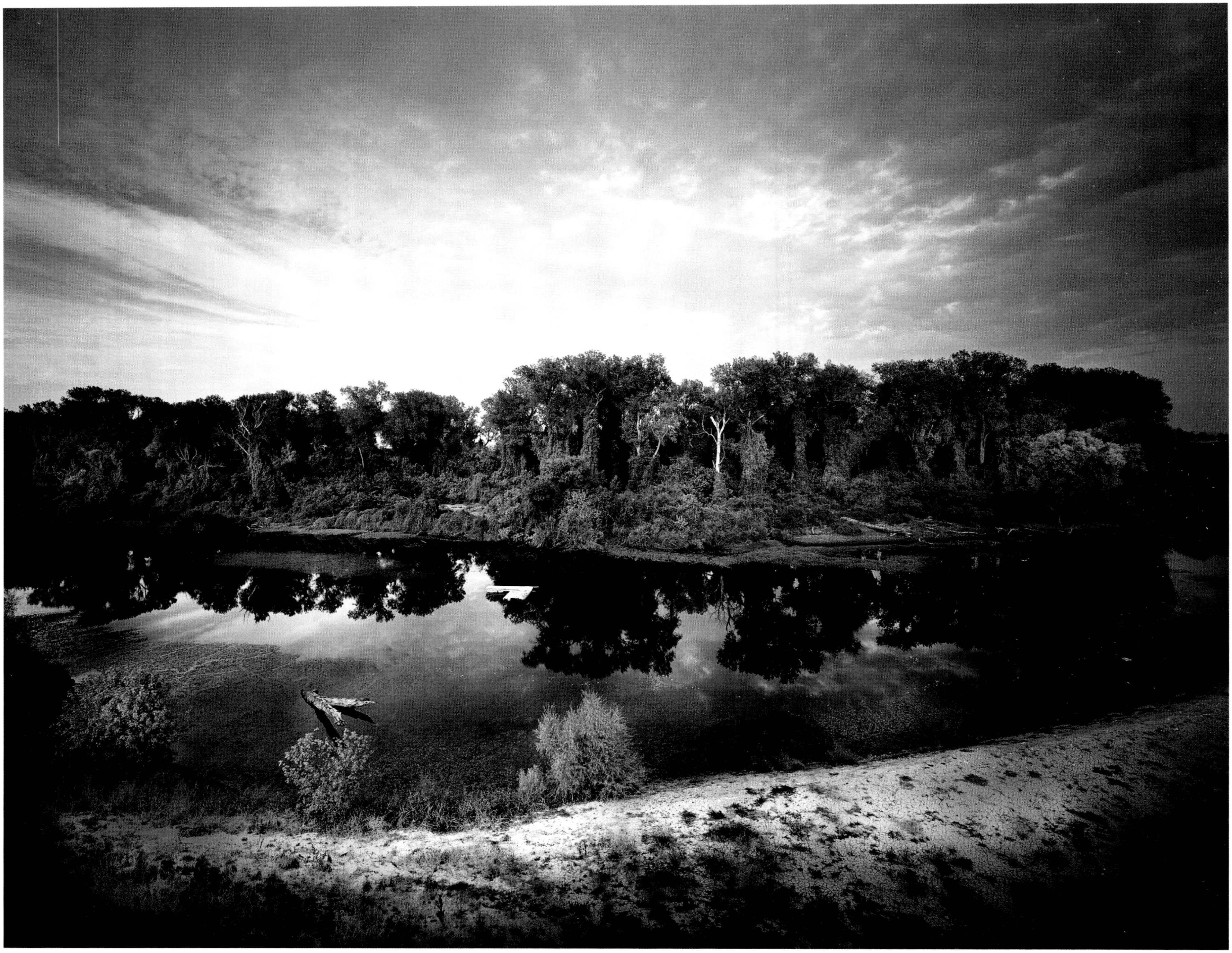

Old-growth riparian forest at Murphy Slough, 1998

SACRAMENT

HOMAGE TO A RIVER

PHOTOGRAPHS BY GEOFF FRICKER

TEXT BY REBECCA LAWTON

FOREWORD BY STACY CEPELLO

Heyday, Berkeley, California

This book was made possible in part by generous funding from the Resources Legacy Fund Foundation, Caroline Burkett in honor of Aline Burkett, and Sierra Nevada Brewing Co.

Library of Congress Cataloging-in-Publication Data

Fricker, Geoffrey, 1947- photographer.
Sacrament : homage to a river / photographs by Geoff Fricker ; text by Rebecca Lawton ; foreword by Stacy Cepello.
pages cm
ISBN 978-1-59714-243-4 (hardcover : alk. paper)
1. Sacramento River (Calif.)--Pictorial works. 2. Sacramento River Watershed (Calif.)--Pictorial works. 3. Black-and-white photography--California--Sacramento River. 4. Sacramento River (Calif.)--Description and travel. 5. Sacramento River Watershed (Calif.)--Description and travel. 6. Sacramento River (Calif.)--Environmental conditions. 7. Sacramento River Watershed (Calif.)--Environmental conditions. I. Lawton, Rebecca, 1954- author. II. Title.
F868.S13F74 2013
979.4'5--dc23
2013008839

Cover Art: Aerial view of meandering Sacramento River, 1997
Book Design: Lorraine Rath

Orders, inquiries, and correspondence should be addressed to:
Heyday
P.O. Box 9145, Berkeley, CA 94709
(510) 549-3564, Fax (510) 549-1889
www.heydaybooks.com

Printed in China by Everbest Printing Co. through Four Colour Imports, Ltd., Louisville, Kentucky

10 9 8 7 6 5 4 3 2 1

To my wife, Sandee—G. F.

For the Stan—R. L.

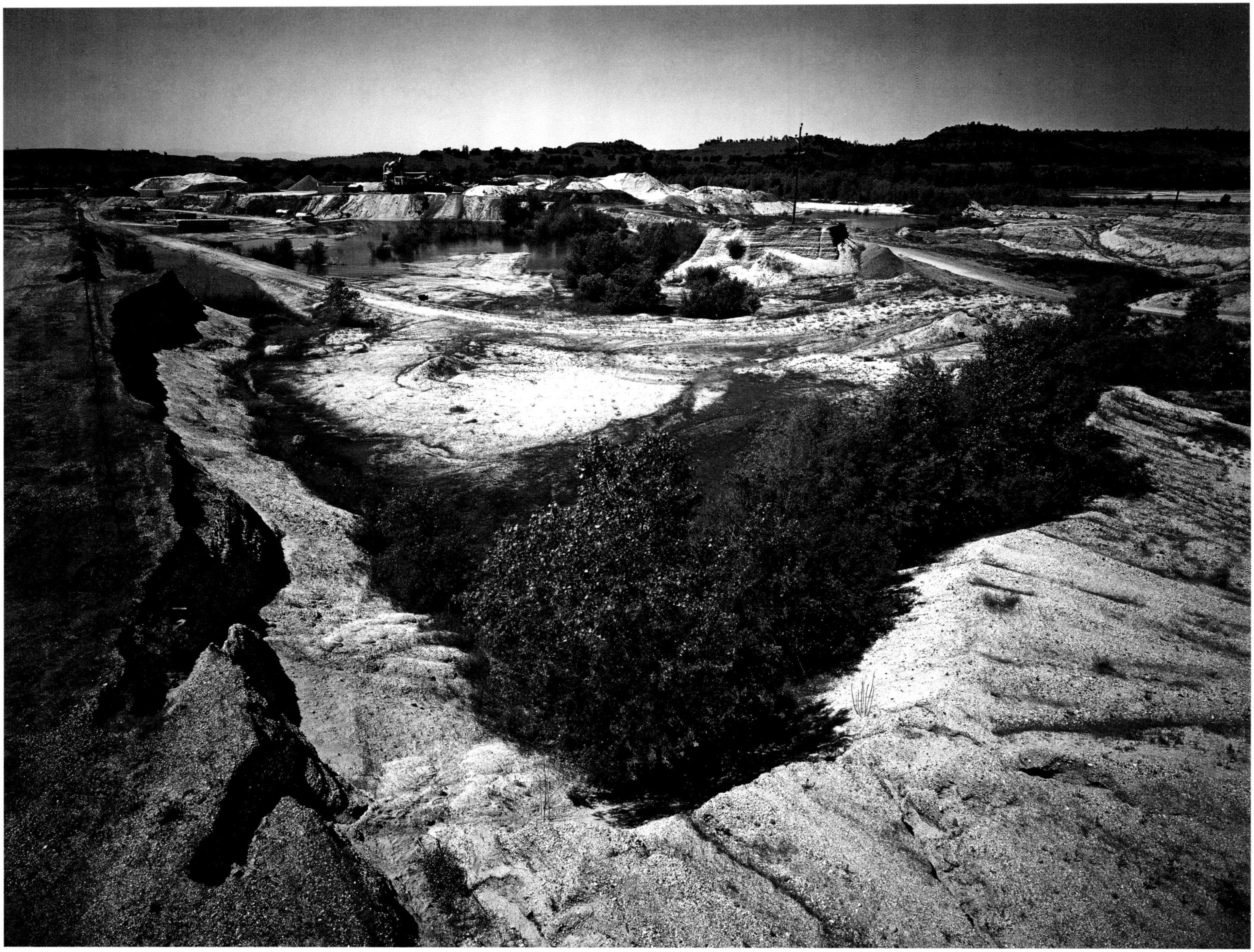

Slickens from Cherokee Hydraulic Mine, 1999

CONTENTS

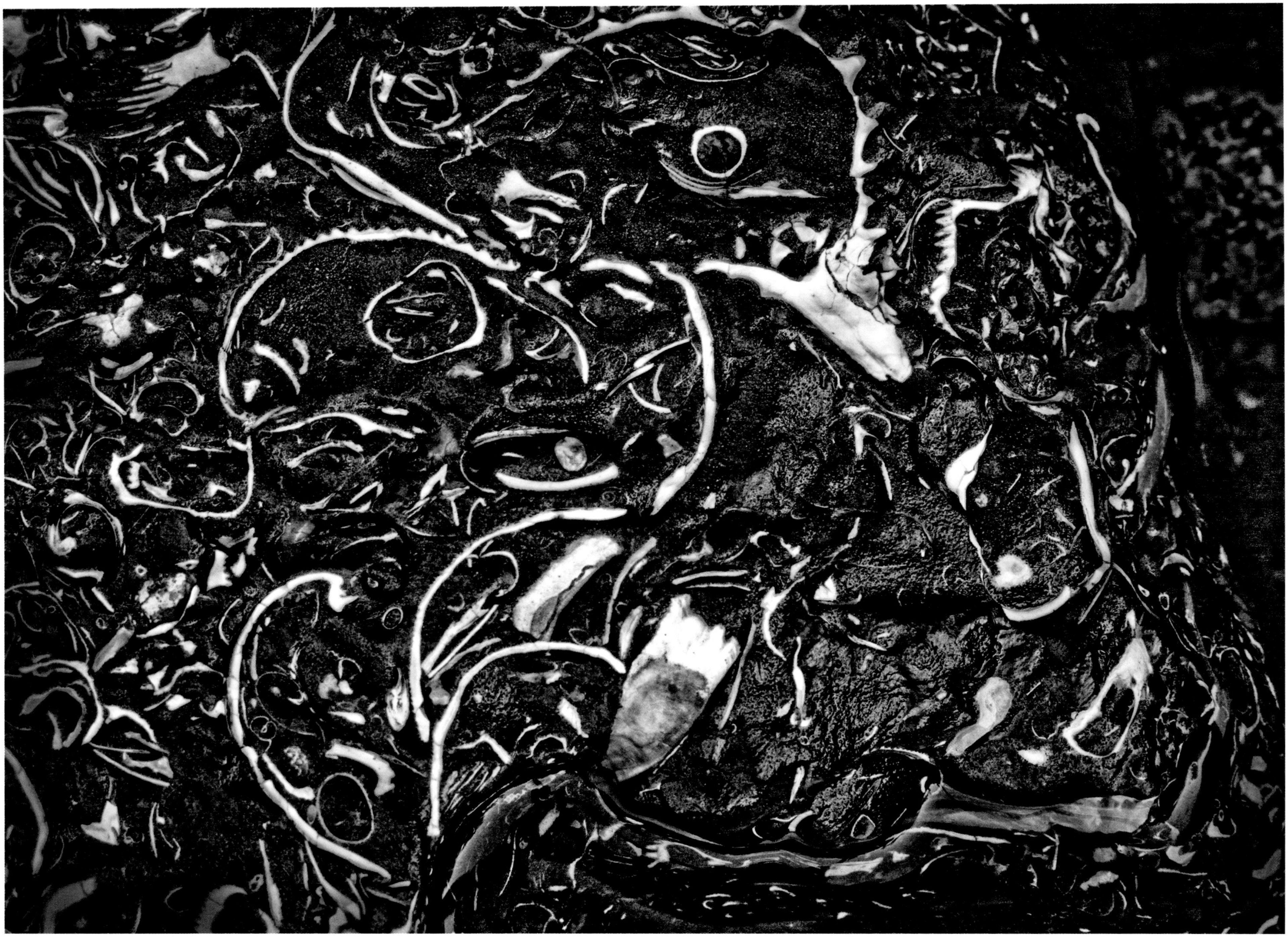

Marine sediment and shell hash from the Chico Formation, 2011

FOREWORD

STACY CEPELLO

GEOFF FRICKER'S PHOTOGRAPHY clearly shows that the Sacramento River is both a natural and a cultural system. Fricker is never content with mere aesthetics in his expansive landscapes: at the heart of each image is an expression of human values—sometimes obvious, but often hidden. I think that even a casual viewer can sense an underlying tension in these artfully and intelligently composed frames. As an environmental scientist, I have spent over thirty years studying and restoring the riverine-riparian ecosystems of the Sacramento River and its tributaries. In Fricker's photographs I see complex interactions between hydrology, biology, and the land—how water carves, erodes, deposits, floods. The human element is, of course, ever-present: here are fishermen and farmers who skirt the boundaries of civilization and wildness. But the real story is the push and pull between dynamic ecosystems and our attempts to simplify, control, and stabilize them.

In the second decade of the twenty-first century—now—Californians will decide the future of the state's water supply infrastructure, the Central Valley's flood management system, and the Sacramento Valley's groundwater policies. In doing so, we will decide the fate of a host of Sacramento River species, including wild salmon, steelhead, and sturgeon; cottonwood riparian forests and ancient oaks; playful river otters; and yellow-billed cuckoos and nesting bank swallows.

Writer Rebecca Lawton invites us to think deeply about what it means to fall in love with a river. Unfortunately, in California it is often a painful experience: death by a thousand cuts as the wildness of our rivers is whittled away. We may not fully understand how to restore large, dynamic ecosystems, but we can no longer deny the long-term damage we've inflicted. "One of the penalties of an ecological education is that one lives alone in a world of wounds," states Aldo Leopold in *Round Valley* (1953). He goes on to say, "An ecologist must either harden his shell and make believe that the consequences of science are none of his business, or he must be the doctor who sees the mark of death in a community that believes itself well and does not want to be told otherwise."

Water is at the much-contested heart of the Central Valley. The need to build ever-taller levees, the need to maximize urban space and farmland, and the need to provide the statistical illusion of lowered flood risk—these are in fundamental opposition to fostering public safety, the diversity of native species, the productivity, and the health of the Sacramento River. Our cumulative efforts to control and stabilize the Sacramento River stifle the fish and wildlife species adapted to and dependent upon its ever-changing nature.

When the river channel meanders freely—without tightly spaced levees, hardened water diversions, and riprap—and is allowed to periodically overflow its banks, there is equilibrium. Floodplains and riverine habitats are created and renewed by erosion and deposition. Great Valley cottonwood forests grow in bands on point bars; colonies of bank swallows burrow into the river's banks to nest and raise their young; fallen trees and natural banks create the shaded channel margins that protect juvenile salmon on their downstream migration.

On the other hand, where the river is squeezed by narrowly spaced levees—built one hundred or more years ago to prevent mining debris from clogging waterways—it runs deep and fast with little or no opportunity to deposit sediment. This lack of equilibrium results in the progressive scouring of the channel and levees, causing the river to become an agent of destruction. The lack of space creates an untenable cycle of bank erosion, maintenance, and repair that will inevitably result in a supersized, rock-lined ditch.

To create a sustainable and resilient flood management system, capable of long-term adaptability to changing hydrological and climatic conditions, we must, as the Dutch government has phrased it in its own flood management plan, provide "Room for the River." A flood system must not only manage risk but also accommodate the river. We can set levees back from the banks where it is possible, in order to allow the river to occupy a portion of its floodplain. In the northern Sacramento Valley there is visible progress toward the coexistence of human needs and natural processes. Through federal and state efforts, supported by nonprofits and under the watchful eye of local governments, flood-prone lands are slowly being acquired and protected for future generations and riparian forests restored. Already we've seen measurable reductions in flood damage and increases in endangered species populations.

We must provide physical space for the Sacramento River through the setting back of levees and the restoration of floodplains in order to insure its future. Perhaps this is best accomplished by providing mental and emotional space for it first. We have the good fortune to have before us this perfect marriage of Fricker's acute visual documentation and Lawton's succinct prose, which provides the context for the imagery. *Sacrament* is both an introduction to the river—its natural and human history—and a glimpse of it through the eyes of an artist. Fricker and Lawton's exceptional collaboration of mind and spirit is sure to lead readers to a deep and abiding appreciation of one of the West's great rivers.

Stacy Cepello is an environmental program manager at the California Department of Water Resources.

PREFACE

GEOFF FRICKER

IT'S IMPOSSIBLE—WELL, PERHAPS not impossible but certainly highly suspect—to approach any river-related topic without a bit of meandering. The motivation and inspiration for this body of work depicting the Sacramento River comes largely from my family and my early years living on a nearby river, the American. Our home was on the bluffs, and I spent hundreds of hours fishing, swimming, and catching crawdads. The salmon's cyclical return from the ocean to reproduce—and, having achieved this, to die—is a magical time, though it brings with it the smell of decay. To this day that smell calls forth the river of my childhood. Naturally I developed a fascination with the history of the human presence on the banks of the river as well.

My connection to rivers, creeks, and water, as well as to photography, runs deep and long. Visual art was a part of my childhood. My mother, Dorothy Brubaker Fricker, painted watery landscapes. She encouraged her children to be creative and to follow their hearts. Her father, Owen Brubaker, was a successful farmer and the ultimate handyman. When Grandpa wasn't working on the ranch, he made accomplished photographs of harvesters and apricot dryers, technological artifacts worn smooth by the touch of human hands.

When I was a young child, Grandpa Brubaker taught me how to build boxes. As simple as they were, they were my introduction to making things with tools. In the 1980s, I designed, and worked with a Japanese craftsman to build, a specialized 8 x 10 camera. I used this camera for most of the images in this book.

Many of these images are taken from an elevated position, in order to give visual context. I retrofitted a hydraulic lift on my pickup truck to achieve this perspective. I think my box camera and this "nuts and bolts" solution for an elevated platform would have tickled my maternal grandfather. Quite a few of my images are about restoring or removing what was created by the technology of earlier times—for example, a dam before its removal, or the addition of fish screens and ladders. Using an elevated perspective, I am able to show the dam relative to the rest of the landscape, revealing the reason for its placement.

My father, Dr. Geoffrey Fricker, a radiologist and photographer, inspired both my interests in science and in using a camera to record my surroundings. I've embraced his belief in science, especially with regard to the health of the rivers. His practice in Sacramento in the early fifties brought our family from Hemet to Fair Oaks and the American River.

Dad is an accomplished photographer. He was a leader in the Sacramento chapter of the Sierra Camera Club and remains active in the Sierra Club today. As a child, I would often fall asleep to the sound of the print washer from the converted washroom/darkroom adjacent to my bedroom. In the 1970s, Dad had a solo exhibit at the San Francisco Museum of Modern Art. At 93, he is still a prolific photographer with his digital camera. My dad gave me my first two cameras. Both my parents instilled in me their love of beauty and knowledge, often taking my siblings and me to the art museums of Sacramento and San Francisco.

When I was a teacher's assistant to John Collier Jr. in my MFA program at the San Francisco Art Institute, I saw through the eyes of a cultural anthropologist. John promoted the use of the camera as a documentary tool. Without being postmodern about it, John understood that cultural artifacts signify meaning. I think we all feel the truth of this subconsciously. The photographer Walker Evans understood this. One of his images of a migrant worker's quarters depicts a wash pan, an oil lamp, and a dirt floor. One immediately understands that this is a residence without plumbing, electricity, and finished floors; it is a home rented to someone with very modest means. Photography can speak volumes about how we live in a place.

In the mid-1970s, I read one of the most game-changing and influential books I have ever read: John Fowles's *The Tree*, which also includes photographs by Frank Horvat. Fowles's introductory essay had a profound effect on me as a photographer, educator, and environmentalist. He argues convincingly that the camera insulates us from full immersion into nature. He identifies our addiction to tools and purpose, which, he argues, estrange us from the true mystery of nature. As a result, I prefer not to photograph mere scenery: instead, I strive to describe visually our human relationship to, or estrangement from, the landscape.

In the 1990s, I traveled with my wife, Sandra, and my two sons to Abergavenny, Wales, my grandfather's birthplace. It turns out that my great-grandfather, Charles Fricker, was a photographer as well. When we inquired about some of the specific locations of his photographs of Wales, handed down to us in an old photo album, we discovered that a local museum housed many of his photographs in their collection. One of his images is still used in a tourism advertisement for the area around Abergavenny. Like in my own work, water is featured in his landscapes—in his case, the river Wye and the river Usk and the barge canals for transporting coal from the mines to Cardiff. An established photographer and bookshop owner, Charles left his business in Wales for new opportunities in the Southern Californian desert. He was to become a farmer.

In 1895, the second-biggest dam in California had just been built on the south fork of the San Jacinto River in Riverside County. Charles purchased some of the water captured by this new dam to nourish his newly acquired and hitherto arid land in Hemet, California. Along with thousands of other immigrants, the Fricker family traveled across the plains and over the Sierra Nevada to reach his new property. Charles wrote detailed journals of his travels:

> Across the lake we entered upon a country curious in the extreme, great stretches of sandy, salt waste with craggy and snow capped mountains in all directions. . . . During the night we climbed 7,000 feet and from sunset for 15 hours we rushed along in an open stretch shut in by mountain peaks. . . . Gradually the mountains gathered in and we were in a gorge of great dreariness, when it became more and more beautiful with a foaming river [Truckee] and pines and snow all around. . . .

> We were in the Sierra Nevada…with snow clad valleys and peaks, all covered with pines, on every hand. And then came the wonderful transformation scene.…Suddenly, as we emerged from the summit and rushed down the other slope, leaving the snow behind, we were in a land of life, rampant and luxuriant, brilliant verdure and glowing color.…Oh! The excitement in our party as we came upon the Californian poppies, the lupines, the orange trees and other sights of that lovely land.

Reading Charles's journals, I was delighted to find that I recognized the territory he describes: as a teenager much of my time was filled with exploring the deeper recesses of this water-carved terrain. With topographical maps in hand, my fishing partner and I sought remote tributaries of the Sacramento River on the west side of the Sierra. We found ferns to wrap our fish, red carpets of ladybugs, and signs of bears, bobcats, and other fauna. We shared Charles's excitement of adventure as we hiked the steep canyons and witnessed headwater arteries flow downstream and converge. Streams of water came together like veins in a leaf joining to create a stem.

My own relationship to the Sacramento is deeply rooted in one of its tributaries, the Feather River. Because of early mining operations, some Feather River water was and still is transferred into Butte Creek, from which our home is a stone's throw away. But the largest percentage of Feather River water is transferred six hundred miles south from Oroville Dam in Northern California to Diamond Valley Reservoir, five miles southeast of Hemet. Diamond Valley Reservoir helps to provide for eighteen million people in the L.A. Basin. It's ironic to me that the water in my backyard comes from a source—the Feather—that is now mostly diverted to the Hemet area, my birthplace; it makes me feel as though my own history is tangled up in that of the river's.

Without much regard to the uniqueness of the rich bioregions of these rivers' original courses, water from the Feather River is whisked away to ever-thirsty coastal California for domestic use. But by far, the bulk of the Sacramento watershed, fed by many other tributaries, ends up in the arid San Joaquin Valley for agricultural use. The end result is, sadly, an impoverished river, riparian corridor, and delta.

There is a finite amount of water with many competing interests. The issues are complex, but the watershed's future is ultimately determined by the number of straws sucking water from it. A thriving agricultural community and environment are dependent on the health of rivers and streams. This questionable future inspired my work in this book.

All of life connects to water and its history. This book is a contemplation of water and our often contradictory connection with it. My sincere hope is that it inspires readers and policy makers to appreciate and protect "a land of life, rampant and luxuriant." Last but not least, I am inspired by the slow but certain path of the river—which meanders in spite of all our attempts to control it.

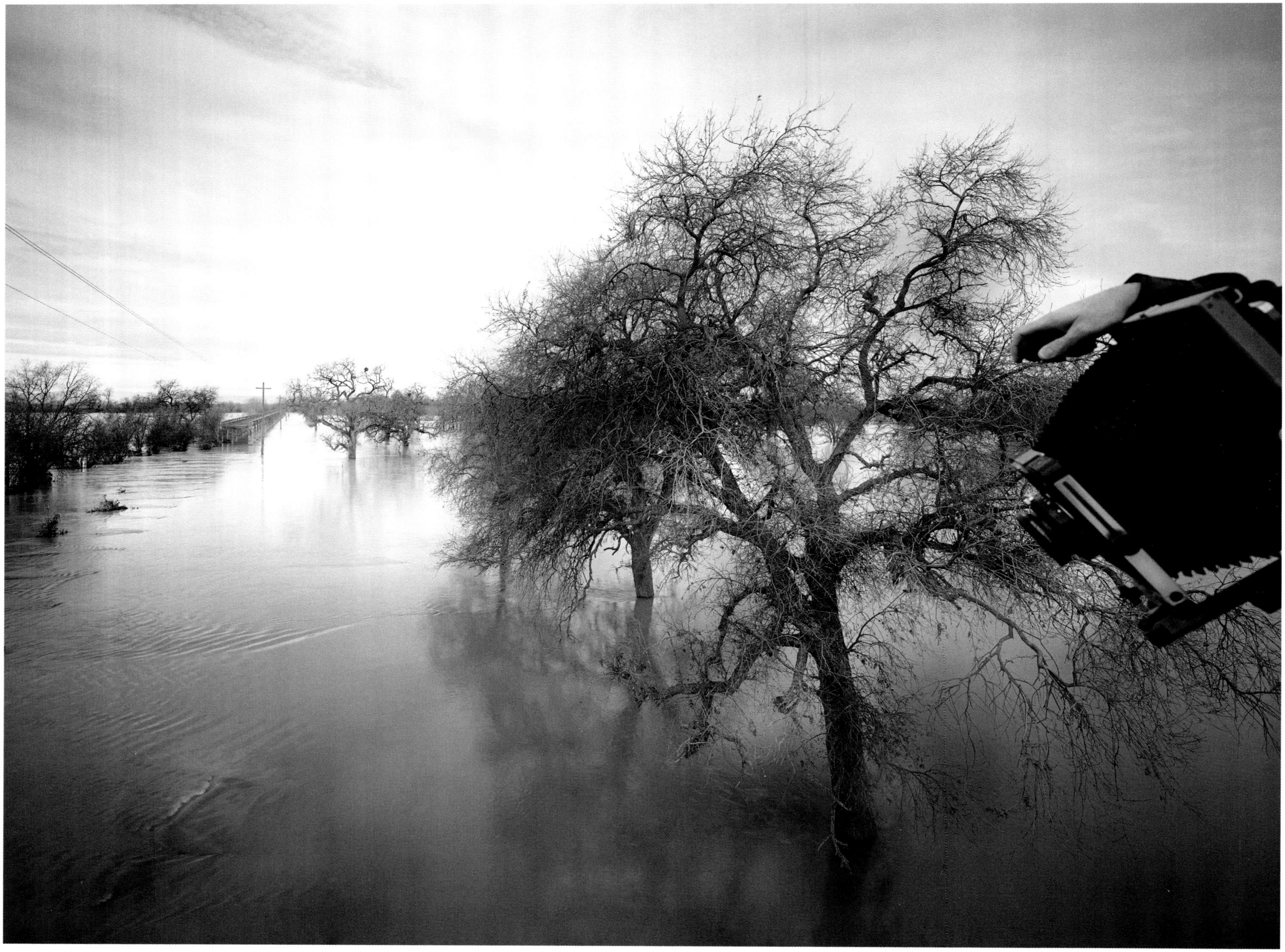

Self-portrait over flooded waters, just south of Sutter Buttes, 1997

ACKNOWLEDGMENTS

THIS BOOK IS AN EXPRESSION of my love for the Sacramento River, but it wouldn't have been possible without the community of people and organizations that has supported my work.

I would like to acknowledge my brother-in-law, Dennis Machida, former executive director of the Tahoe Conservancy. He helped me to dream up this book many years ago. I appreciate Dennis for believing in my work, for discussing it with me for countless hours, and for arranging my first meeting with Michael Mantell of Resources Law Group and their affiliate, Resources Legacy Fund. Michael, thank you for agreeing to look at my work and for your pivotal role in making this book a reality. Two private individuals provided additional financial support: Ken Grossman, owner of Sierra Nevada Brewing Company, and Caroline Burkett, water advocate, fellow artist, and dear friend.

Rebecca Lawton describes the beauty and power of the "Sac," as well as the issues that surround it, so eloquently. I want to express my heartfelt appreciation for her involvement in this project. Her writing captures what the river says to me, and I am honored and humbled by it.

The foreword was written by Stacy Cepello, a systems scientist with the Department of Water Resources. Over the years, Stacy and I have exchanged ideas about science and policy, and he provided opportunities to photograph many parts of the river. His river-related expertise, which includes a deep knowledge of infrastructure, systems, and management, is unparalleled. I was honored when he agreed to write the foreword to the book.

River Partners, a nonprofit river restoration organization, provided fiduciary oversight of the funds. In particular I want to thank John Carlon, executive director, for his support, expertise, and work. John's conversations with Rebecca Lawton provided invaluable perspective for framing some of the key messages of the book.

Over the past thirty years, I have been inspired by John Merz, executive director of the Sacramento River Preservation Trust.

I want to express my appreciation for my colleagues at Butte College who took over some of my teaching responsibilities while I worked on the book. Additionally, I want to thank Clancy Gehrke, Matt Kidder, Wayne Wade, Neal Snidow, Dave Cooper, Marianne Werner, Byron Wolfe, Alex O'Neil, Kathleen Kelly, Kathleen McPartland, Susan Wooldridge, Carol Perkins, Thomas Taylor, Allen Harthorn, Lloyd Andrade, Jeff Howell, Phil Sharrow, Dan Gesick, Chris Kerston, Daniel Donnelly, John Clark, Patsy Ceek, and Adam Henderson for their support and advice. Thanks also to my students whose own struggles to clarify their purposes encouraged me to think deeply about my own. My thanks to all my fellow members of Water in the West. You have all inspired me with your interest in and commitment to issues provoked by water scarcity. Thank you Wendy Brouws for your abiding support for this publication and for helping me understand the publishing world.

Malcolm Margolin, publisher of Heyday, took the time to review my work and hear my message. His vision and faith have shaped this book. I have heard it said that "Heyday does books for all the right reasons," and I feel honored to be a part of this project. Malcolm's staff, in particular Gayle Wattawa, provided the best, most talented support that I could have ever imagined—prompt, professional, and positive. What a privilege it has been to work with Heyday.

A small army of friends and colleagues has helped me on different parts of this journey. I want to thank several former students: Davin Schreindl, for the many hours of work and technical support; Ben North, for being my hiking and photographic partner; and Daniel Barth, for sharing his love of history with me. Thanks also to Ken Fleming and Tim Calhoon for bringing me to special places in the mountains and to Paul and Susan Minasian for helping me to understand different perspectives surrounding water.

Finally, I want to thank my family. I want to express my love and gratitude for my sons, Andrew and Joel, who have been my travel buddies on many of my outings. I want to thank my parents, Dr. Geoffrey Fricker and Dorothy Brubaker Fricker, for their unfailing love and support of everything that I have done. There are so many other family members who have encouraged me, chatted with me, attended my photo openings, and humored me when I needed to make yet another photograph. My beloved family includes my sisters, Ann Fricker Sandberg and Jen Fricker; my in-laws, Miyako and Hideo Machida; Robert Skip Sandberg; and Connie and Mel Gon. Lastly, I want to thank my wife, Sandee Machida, who supported me, accompanied me, edited my work, took care of the family and household while I was away on photo excursions, prepared my lunches, and even considered sending out rescue patrols to get me.—G. F.

HEARTFELT THANKS TO Geoff and Sandee Fricker for sharing your time and home with such kind hospitality. Your creativity in not only envisioning this book but also designing a life by the river inspires me. Gratitude to Stacy Cepello and Adam Henderson of the California Department of Water Resources for giving generously of your deep expertise and time on the water; John Carlon of River Partners and John Merz of Sacramento River Preservation Trust for sharing your passion for and wide knowledge of habitat preservation and restoration; Malcolm Margolin and Gayle Wattawa of Heyday for your enduring faith and creative vision; Sister Michaela of Sky Farm, Joe Medeiros, and Lillian Howan, for your support and intelligent insight during various stages of this project.

And mostly to Rose McMackin, who has inherited my love of the river, and Paul Christopulos, who insisted that I write about it.—R. L.

S A C R A M E N T

REBECCA LAWTON

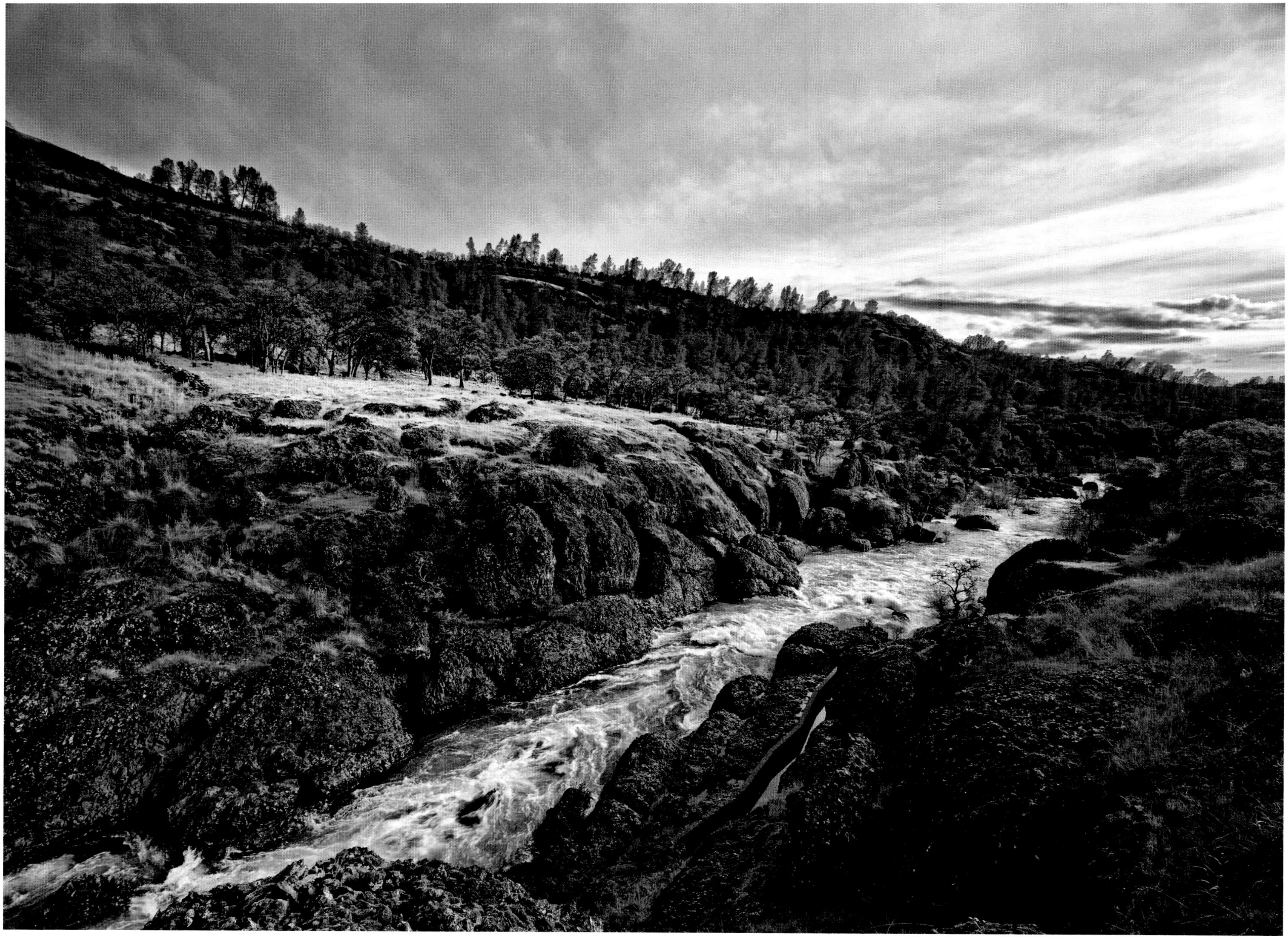

Looking west from Bear Hole on Chico Creek, 2011

HEADWATERS

TO FALL IN LOVE WITH A WILD RIVER is to be changed forever, heart and soul. To fall for a river in California is to live with the scars and ghosts of loss.

El Rio del Sacramento is a river christened in reverence for divine grace. Inspired by a tradition of miracles and redemption, Gabriel Moraga's Spanish expedition baptized the river on a day of Holy Sacrament in 1808. Today we still use that same faith-given name: *Sacramento—river of sacrament, devotion, allegiance, reparation.* Moraga's party was neither the first to encounter the great water nor to venerate it—the Sacramento had been a blessing to indigenous bands of people for a dozen millennia. It was the life-giving stream they hunted and fished in winter; its cool corridors drew them into the hills in summer alongside migrating game and salmon. Many river devotees today affectionately call it the *Sac:* the mighty Sac, the fecund Sac, the productive, amazing, abundant Sac.

We love the Sac not only for what we see in it now but also for its past life and future potential.

Given its imposing volume of water and sheer size of its drainage, the Sac is a great river. Its *watershed,* the land from which all water above *and* below ground drains to the same stream or ocean, covers 27,200 square miles—almost 17 percent of California's total area. It is one system, encompassing a network of tributary waterways feeding the whole. *Tributary* has roots in the fourteenth-century Latin *tribus,* or tribe. Clan, group, collective, race, folk, nation, family. The main Sac comprises a family of streams that include the Pit River, McCloud River, Mill Creek, Deer Creek, Butte Creek, Big Chico Creek, Battle Creek, Feather River, and American River from the east; Clear Creek, Cottonwood Creek, Stony Creek, Cache Creek, and Putah Creek from the west. In the way of families, the members are joined in an association with unifying and defining characteristics while possessing unique attributes that set them apart.

The tributaries flow from the state's north-central mountain ranges and plateaus, collecting 33 percent of the state's total runoff before reaching the Pacific Ocean. Originating near the base of Mount Shasta where the Upper Sacramento, McCloud, and Pit Rivers unite to form the river proper, the Sac has both distinctive volcanic terrain headwaters and a low-lying valley run to the sea. From Shasta Reservoir, which subsumed the Upper Sac, McCloud, and Pit, the Sac flows southward more than three hundred miles to its confluence with California's second great river, the San Joaquin, about thirty miles east of San Francisco Bay. The two mighty waterways join at the *delta,* a wedge of fertile, low-lying land shaped roughly like the Greek letter Δ—narrow at its upstream apex, wide at its generous mouth. The rivers built the delta out of fine sediment deposited patiently, particle by particle, into the rising seawaters of a newly forming bay as the last ice age ended.

You can feel the roots of an ancient world on the Sac. Aquatic populations include remnant runs of fascinating and curious ocean-going species: steelhead trout, green sturgeon, white sturgeon, river lamprey, Pacific lamprey, and four runs of Chinook salmon (in winter, spring, fall, and late fall). Millions of migrating waterfowl stop over on the Pacific Flyway between Alaska and Patagonia every year, inspiring awe not just for their beauty but because they carry forward from an even greater migration in older eras, when the river's diverse habitat seemed never to end. To this day, in some reaches the Sac still supports a streamside world resembling the one the first immigrants must have encountered—deep, wooded, and diverse. In the one-hundred-plus miles of river extending from Cottonwood Creek north of Red Bluff south nearly to Colusa, you can spot eagles, osprey, bank swallows, wood ducks, mergansers, acorn woodpeckers, and other wild birds of the riparian corridor. Beaver and otter ply the river's surface, trailing ripples in their wake. You can stand on the bank and listen to the age-old sound of current sluicing through shallows. You can pick up the earthy scent of mud entrained in high water during winter storms, and the sweet aroma of willow hanging over still pools on hot summer days. You can swim the Sac, boat the Sac, fish the Sac, feel the silt in your teeth when a breeze blows up the Sac. You can laze on a sandbar counting the geese winging overhead against the wide, blue sky as people have done for thousands of years.

Look closer, and you may discern artifacts of transformation as well: deeply cut riverbanks, centenarian hand-built levees, flooded roads, broken flumes, fallen bridges, and empty homes. People have not always understood the river and its ways: our unawareness shows in what we've left behind.

Being a great river in the arid American West, the Sac bears on its back the needs of a growing population. In a regime where only twenty to forty inches of rain and snow fall annually, most of the Sac has been altered to divert and store its water. Change is inscribed everywhere on the river's beds and banks, evidence of pressures that mantle California's naturally lovely landscape. The Sac's precious snowmelt rushes to cities through hand-cut tunnels and canals; her winter rains funnel to reservoirs to be held through the long, dry summers.

To the observant, alteration shows in the river's much-reduced streamflow, its displaced and eroding beds and channel banks, the grids of roads and canals imposed on adjacent lands, and the bafflingly complex concrete structures that often endure long beyond their service. These marks of human influence accompany the river on much of its passage from Shasta to San Francisco Bay: no more the natural dendritic system of its origins, the Sac has become a grid of linear ditch and aqueduct.

In its wild state, the river is defined by *meanders, riffles, point bars,* and *pools.* The vocabulary of its untamed character includes *habitat, riparian forest, canopy,* and *succession*. Couched in a peculiar language, the modern river encompasses engineered *reservoirs, bypasses, canals,* and *dams*. Its human-designed world comprises *calculated withdrawals, acre-feet, appropriations,* and *obligations.* Now, too, a modern language of restoration and renewal is emerging: *indicators, revegetation, monitoring, modeling, resiliency. Metrics, recovery, targets, success.* The Sac knows change and the mark of human industry, it's true, but it also knows choice, beauty, and possibility. Along every bend of its seaward path, the Sac tells a fully Western story.

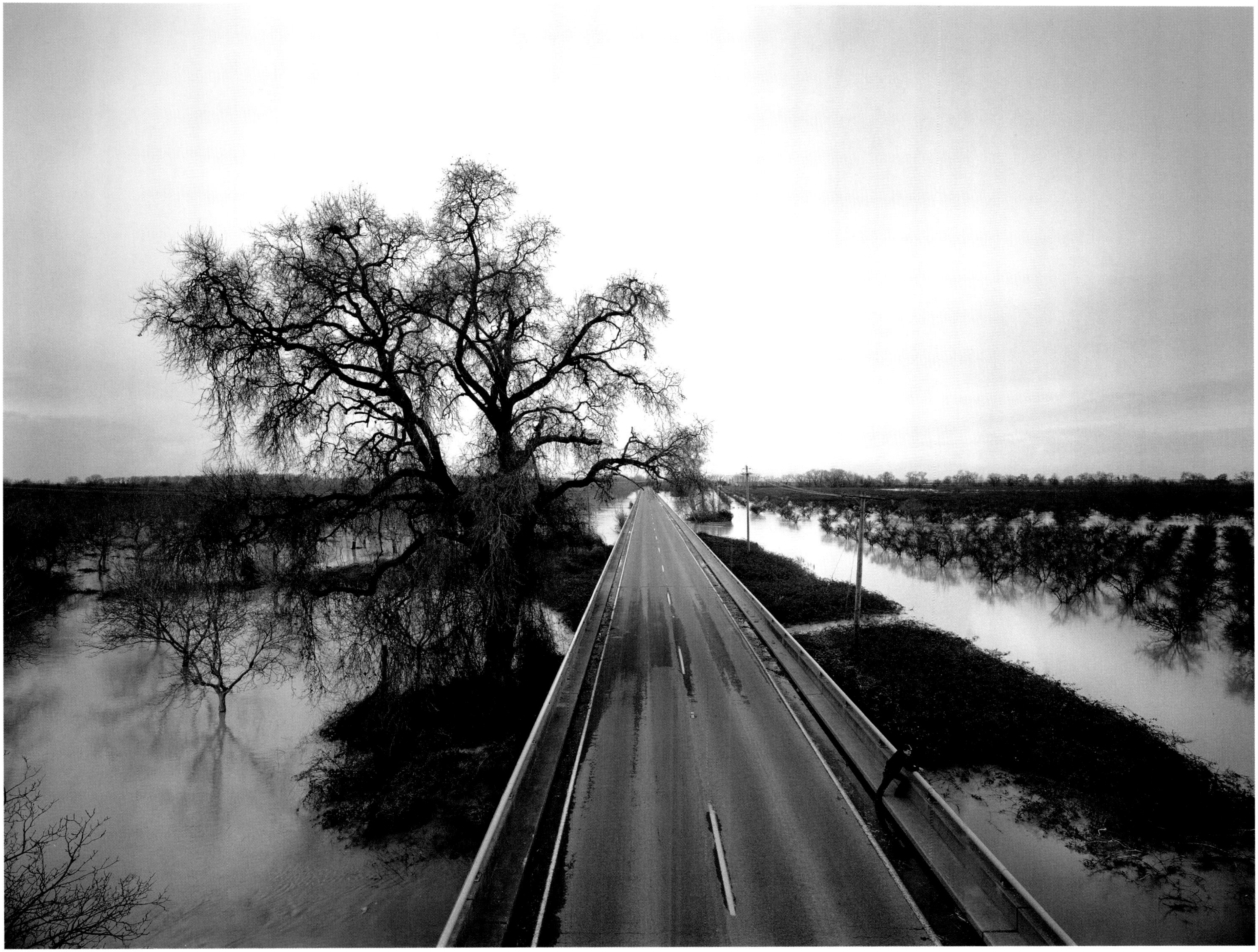

Looking west down Highway 162 from Butte City, 1997

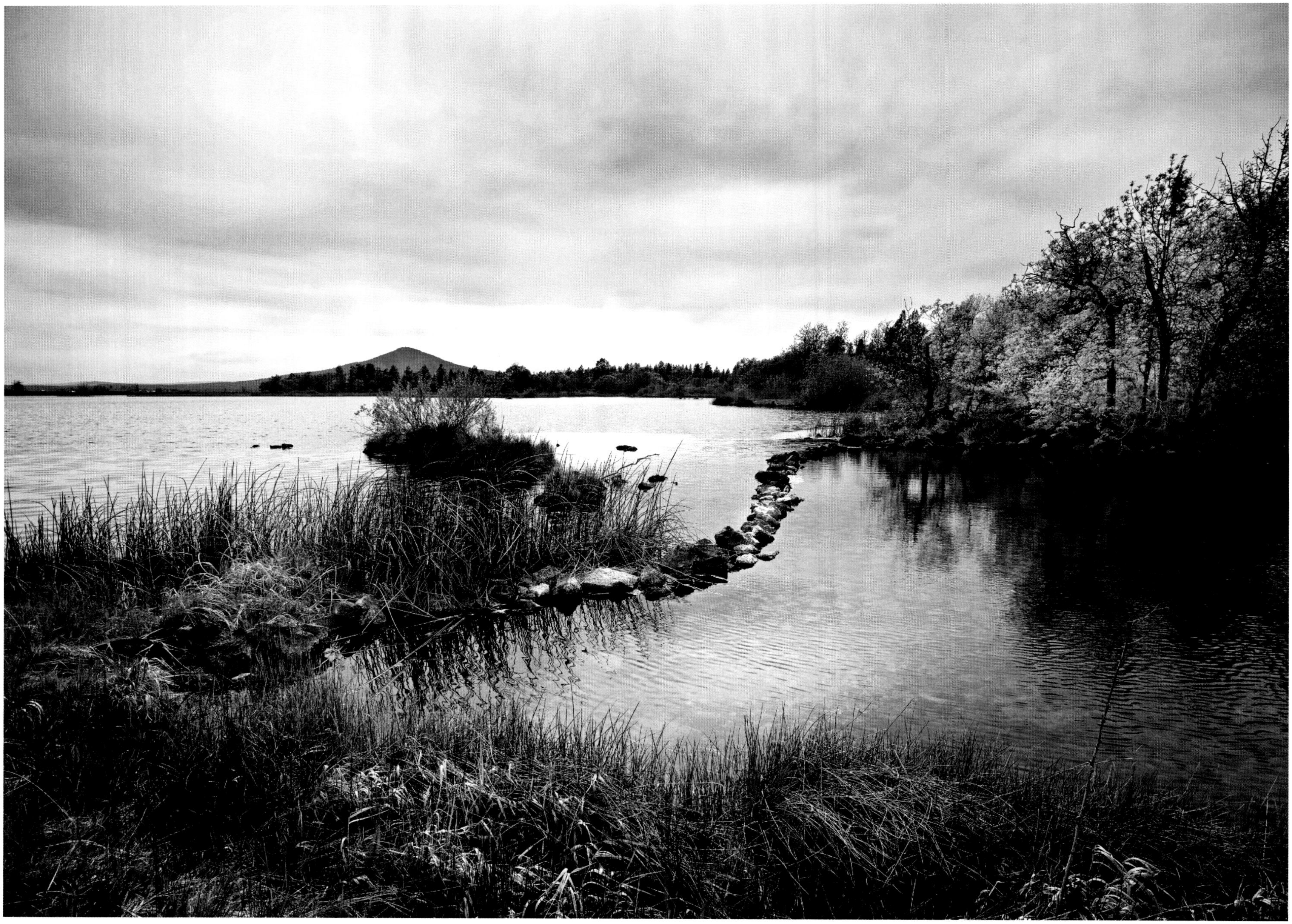

Dam at the river's headwaters used for centuries by the Achumawi to corral the Sacramento River sucker, 2010

Levee repair barge in the delta near Antioch, 2012 »

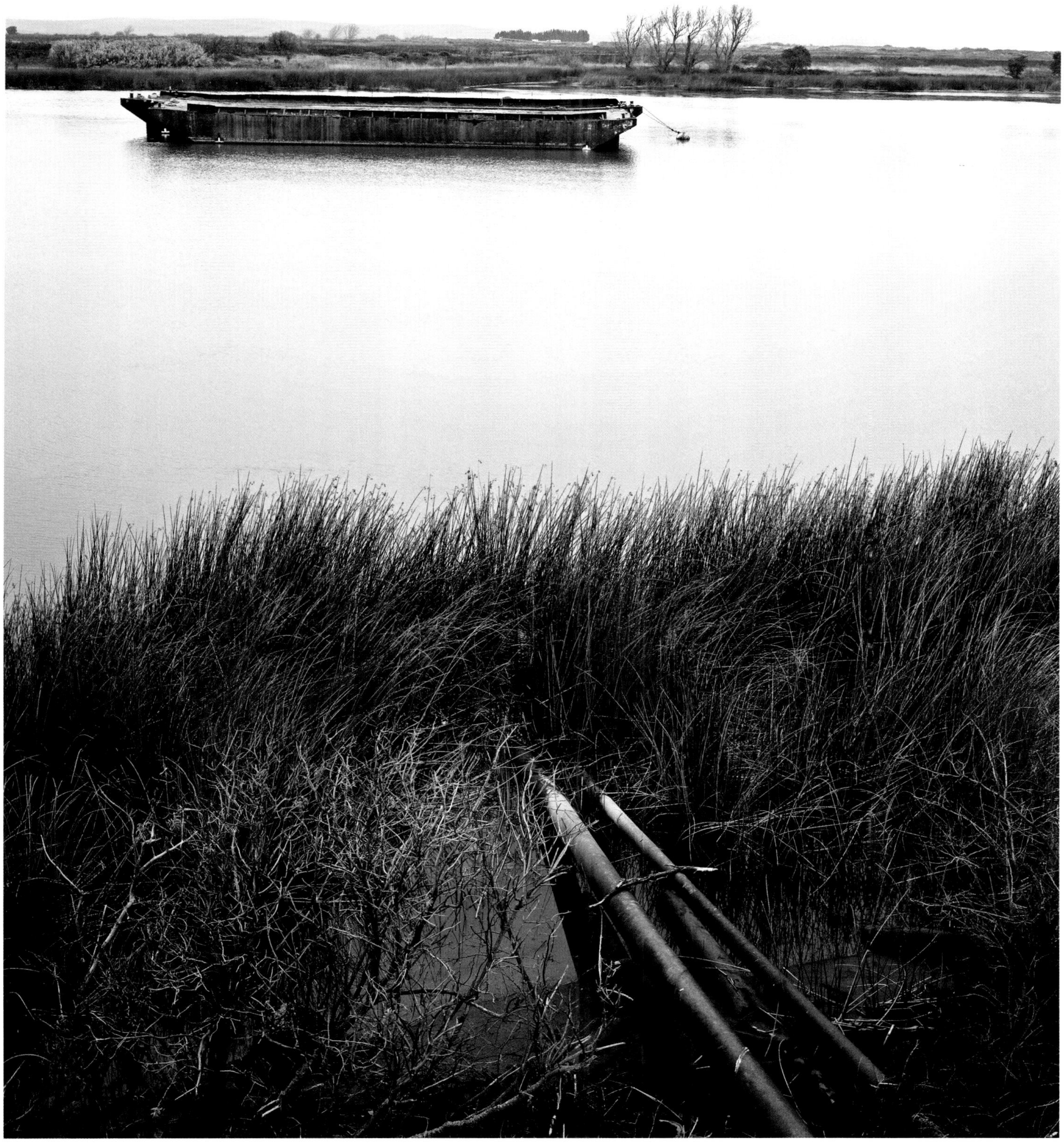

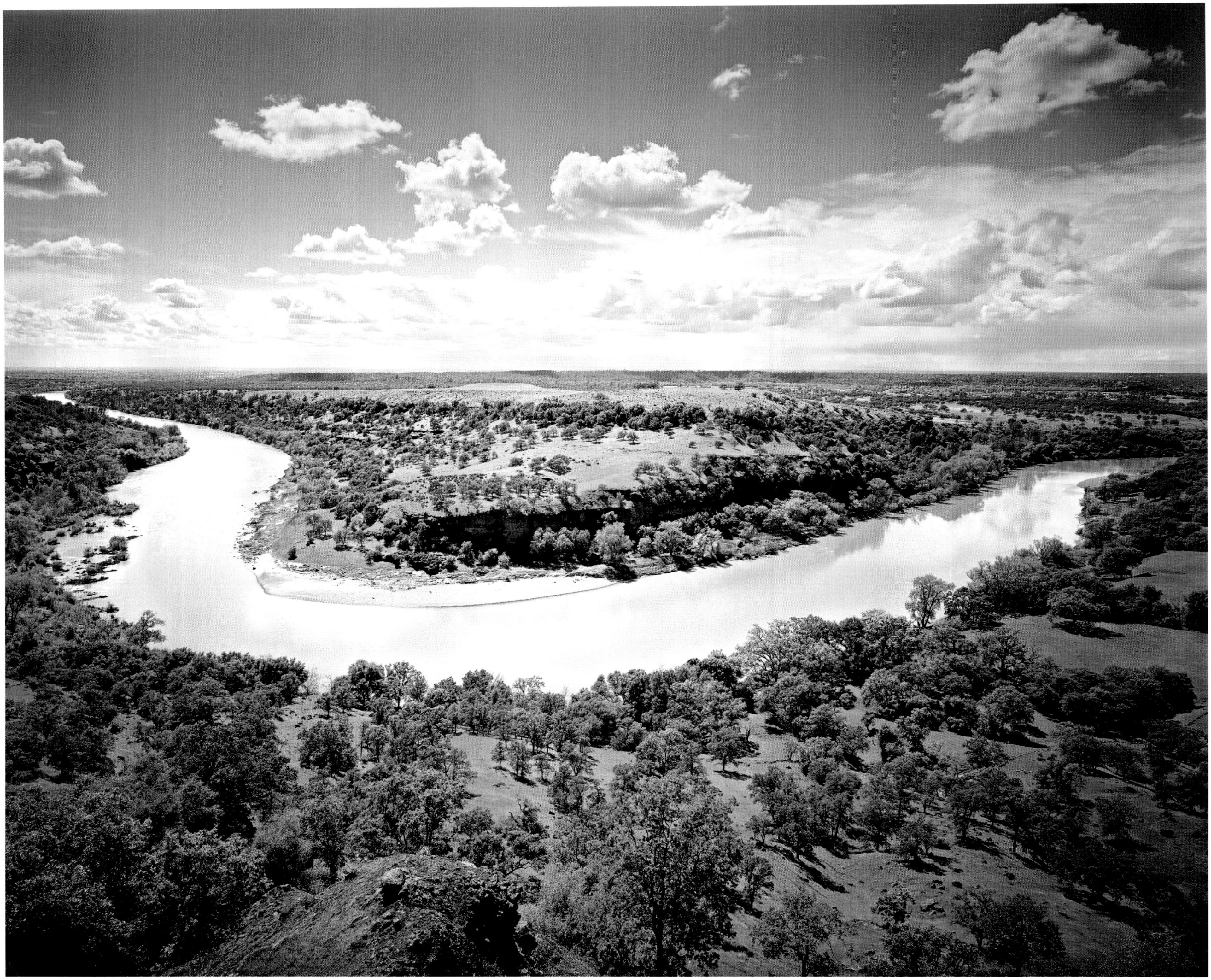

Jed's Point, where Jedediah Smith first sighted the Sacramento River, 2004

Sculpted granite, Feather River, 1995 »

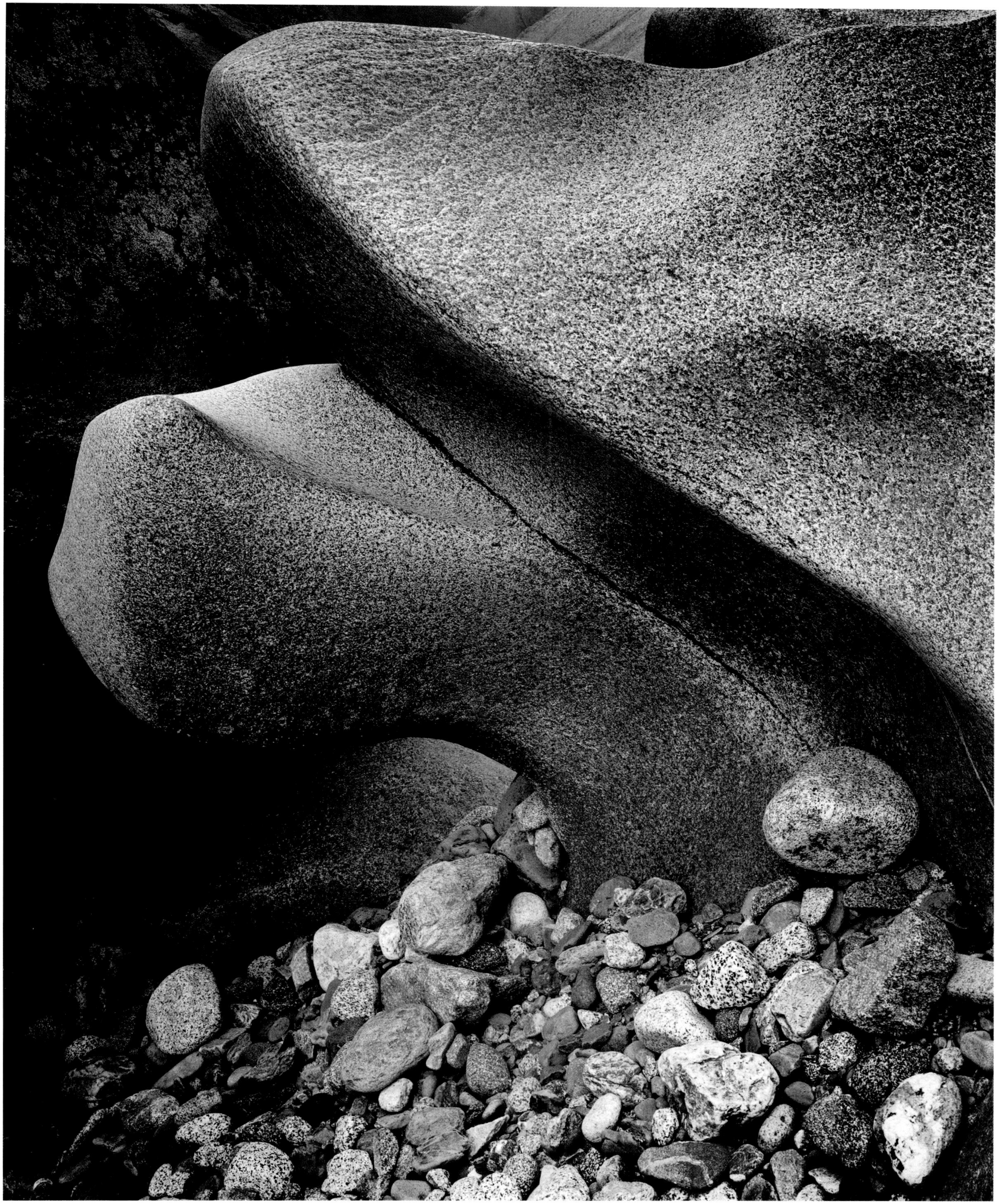

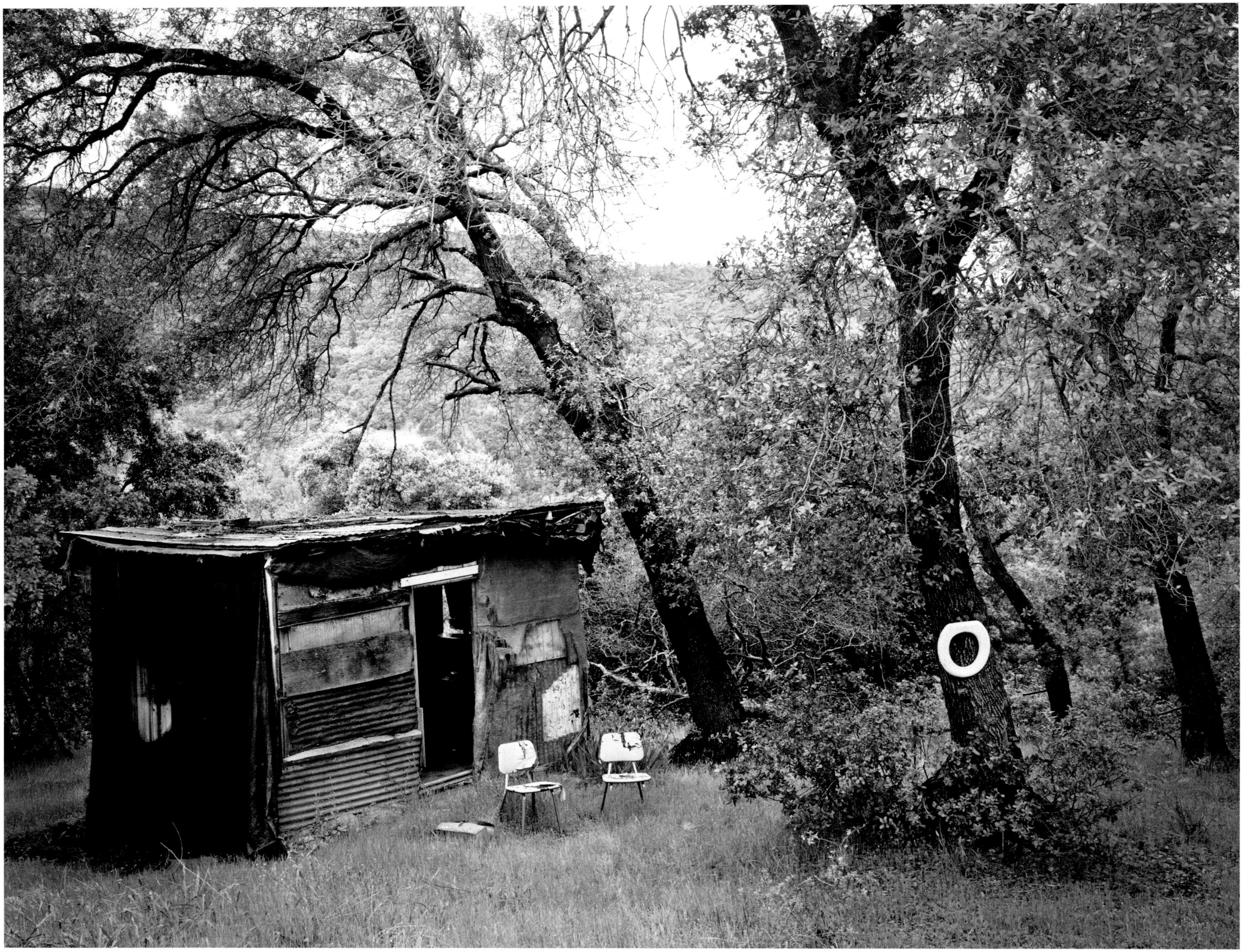

Miner’s shelter, Butte Creek, 1997

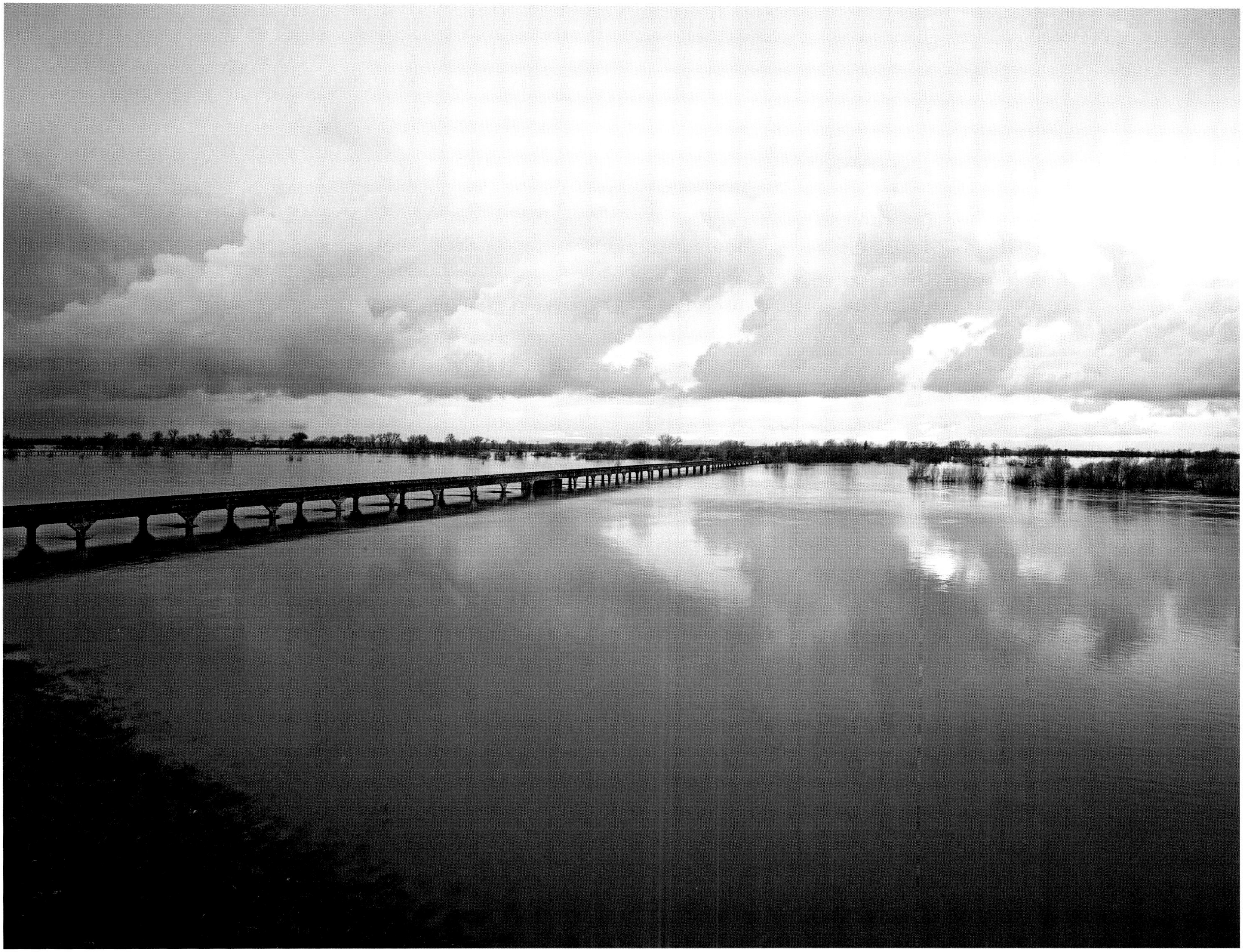

Abandoned railroad causeway at the north end of the Sutter Bypass, 1997

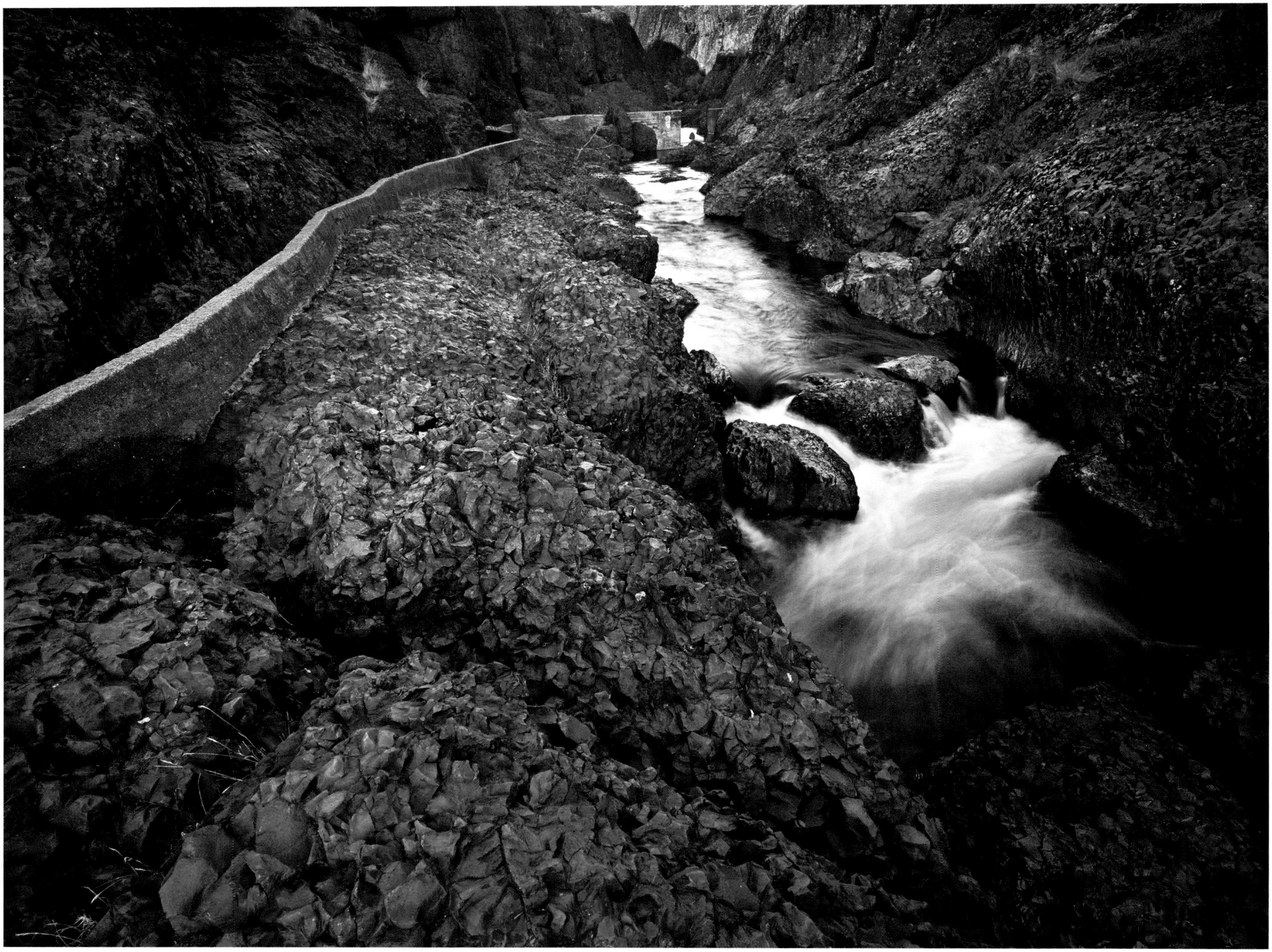

Chico Creek diversion dam and flume, 2005

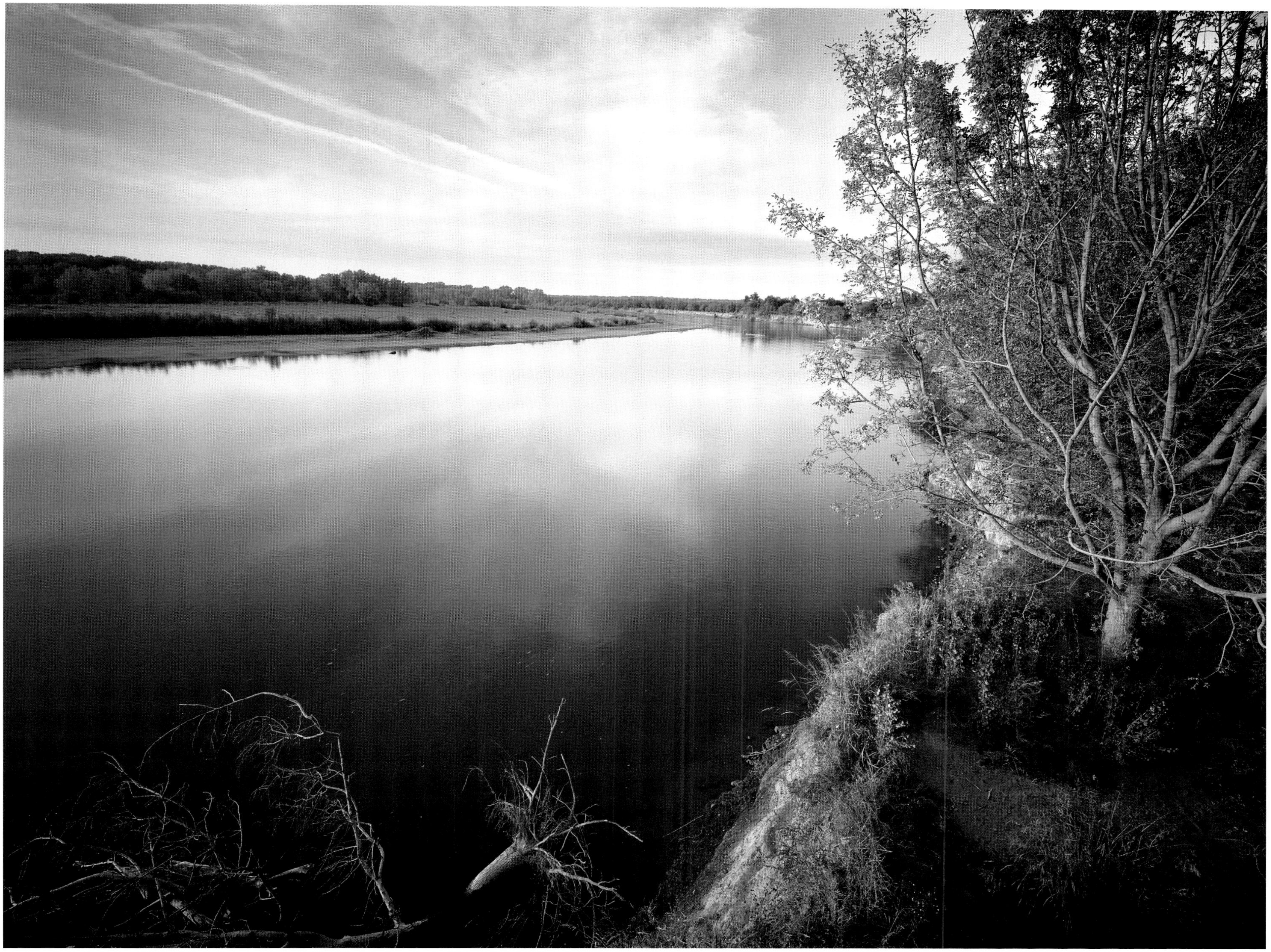

Fallen walnut tree, illustrating how natural processes like scouring and erosion make the Sacramento River's floodplain unstable for agriculture, 1997

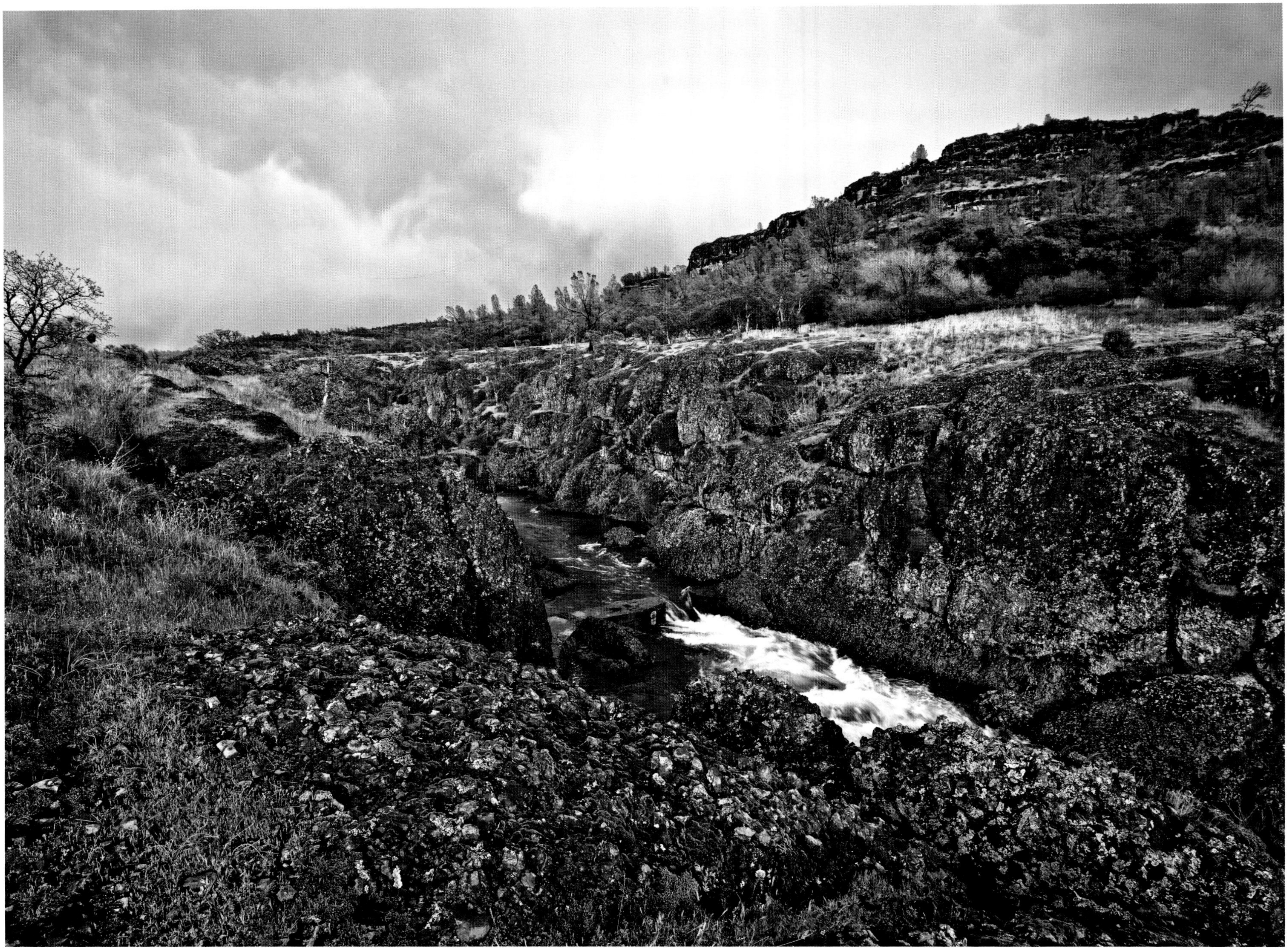

Looking east from Bear Hole on Chico Creek, 2011

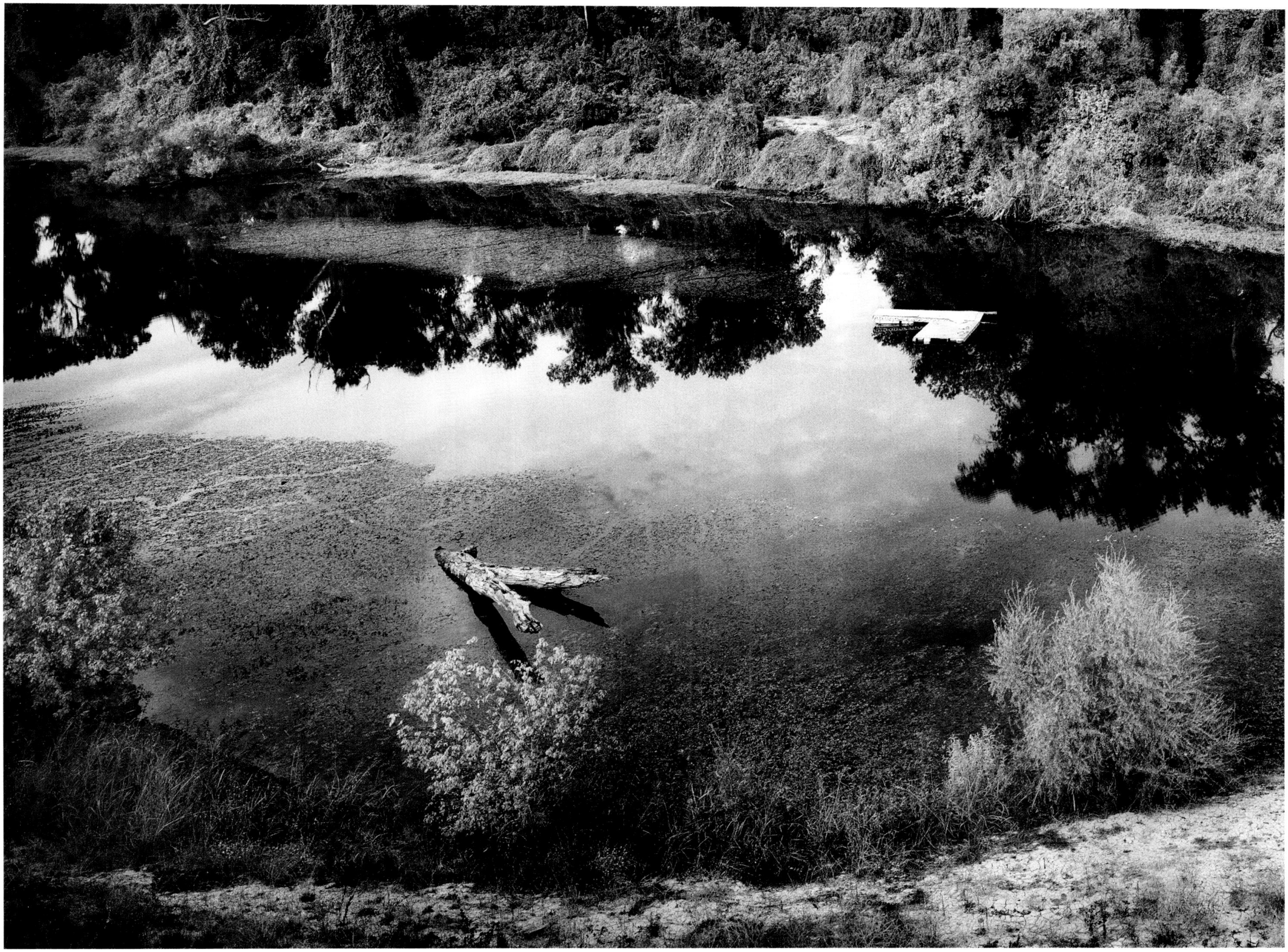

Free-floating dock and submerged cottonwood at Murphy Slough, 1998

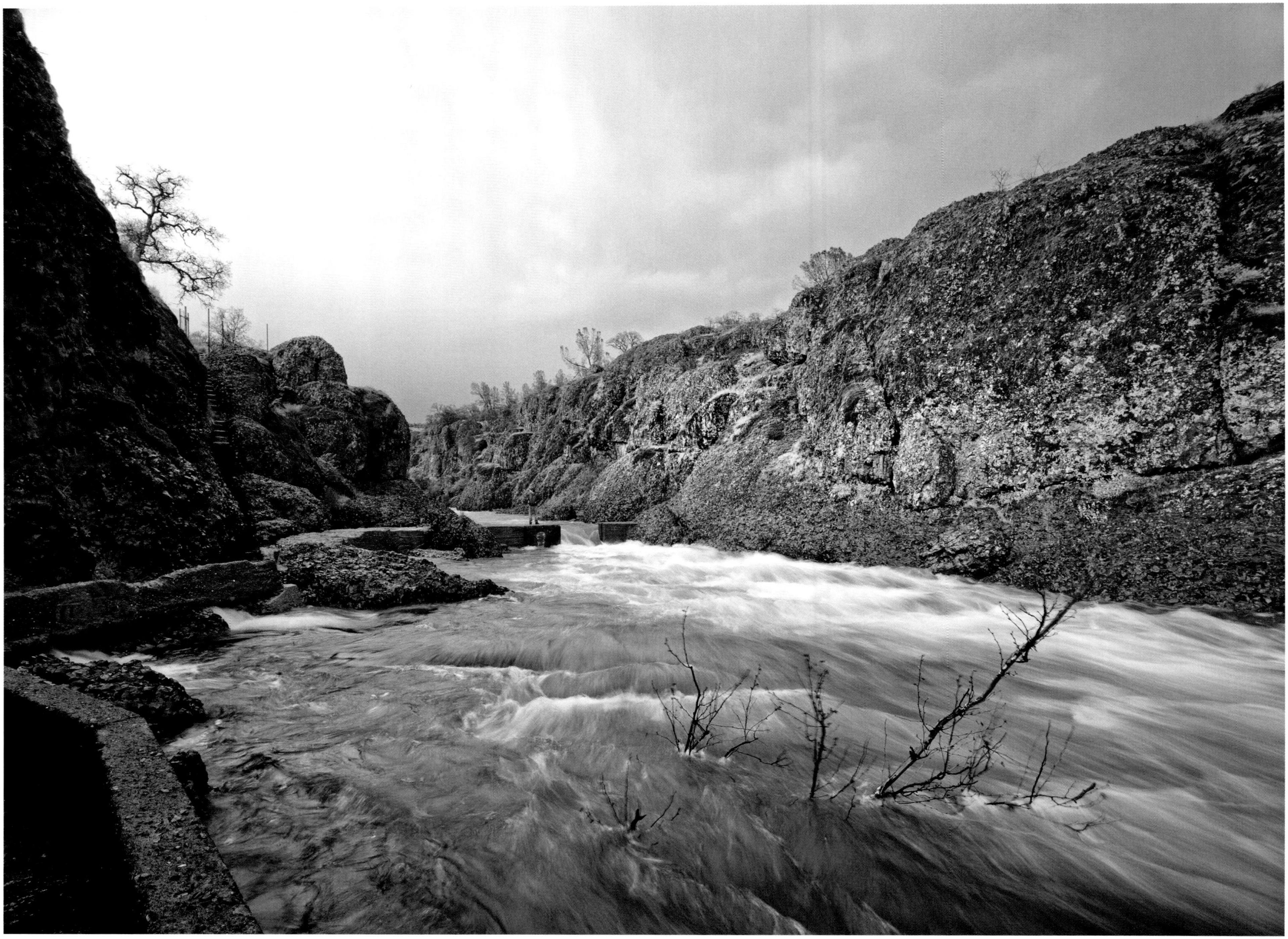

Old diversion dam and flume on Big Chico Creek during high water, 2011

Orchard patterns near Colusa, 2001 »

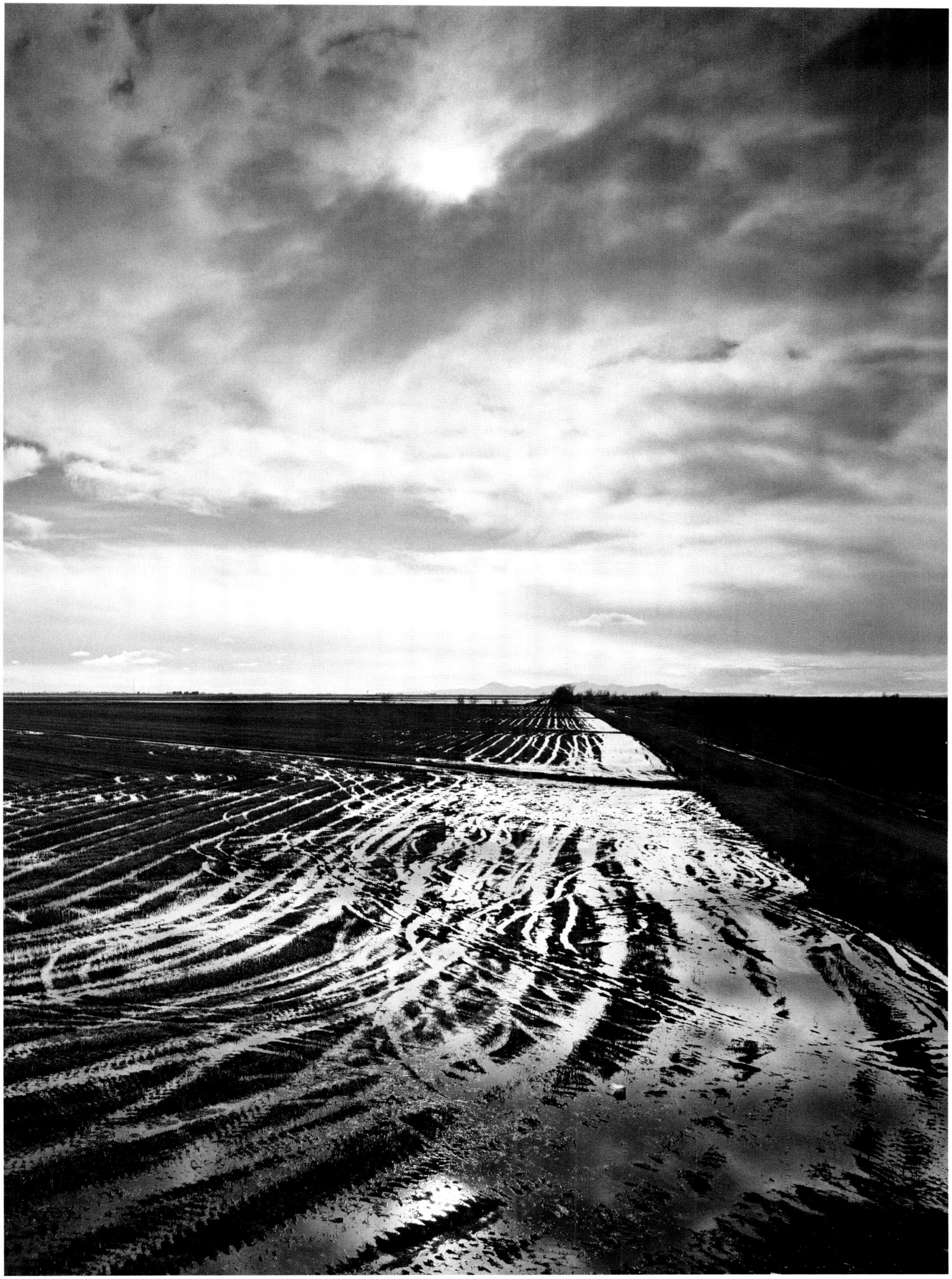

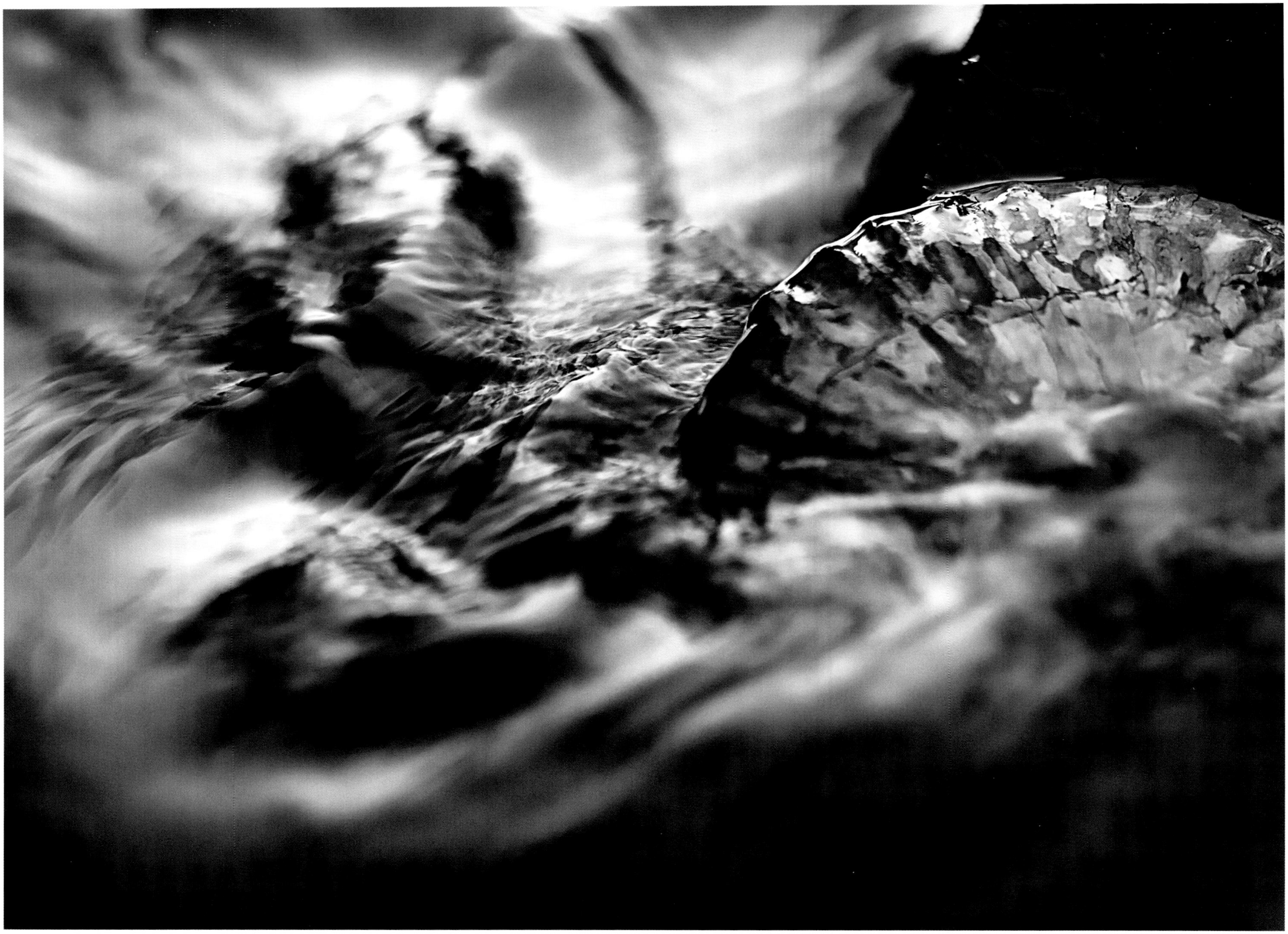

Helical flow over a one-hundred-million-year-old ammonite fossil, Chico Creek, 2011

W A T E R S H E D

THE GREATEST THREAT to a living river is ignorance: inattention to water's innate character and unawareness of the stream corridor as cipher of change. We can no more subdue water than we can stop the tides; we try anyway, and the river responds in highly visible ways.

Water in a river travels a helical path downstream, in a spinning, turbulent, fluid flow. Helical flow determines river's course through its watershed. Water corkscrews into one riverbank then swings back toward the other, spinning all the way. Its circular flow carves bends and enlarges looping turns as it goes. Water has to wander. It's not that it's messy, as may be assumed during floods; it's simply driven to meander by its helical nature.

Water also seeks the path of least resistance; it takes the easiest route. On its way to the ocean, where it is destined to go, it erodes weaknesses in bedrock, concrete, levee soil—anything it can pry into—until eventually it wears them down. It may take millennia for water to progress along its chosen path, given the great durability of some obstacles standing in its way, but it never quits.

Water insists.

Together with a river's hydraulic properties, its geological terrain determines a watershed's character. California's Central Valley is basically a trough-shaped basin into which a deep, varied mass of sediment has been deposited. Before there was a fertile valley with the Sac winding through it, there were uplands from which mineral-rich, abundant volumes of soil were carried. Throughout geologic time, great mountain-building pressures, resulting from the collision of continents and expansion of ocean floors, pushed up the ancestral highlands of California's northern mountains, range after range.

The valley's gradually deepening sediment diversified as the mountains evolved. Light-colored, quartz-rich alluvium washed down from the granite slopes of the hundred-million-year-old predecessors of the Klamath Mountains and Sierra Nevada. As the center of mountain building shifted, lava extrusions from the volcanic twenty-five-million-year-old Cascade Mountains and Modoc Plateau contributed dark, basaltic material to the valley's accumulating deposits. When the ancient Klamath Mountains and Sierra Nevada got their second wind for mountain building, more material sluiced into the valley from each range, beginning some 7.5 million years ago from the modern Klamath Mountains, followed four million years later from the new iteration of the Sierra.

Those initial sediment-depositing events didn't even include the younger material that had yet to carry in from the west. As the Coast Ranges rose, three million years young and counting, more sediment gathered deeply in an elongate valley floor nestled between mountains.

The Sac carved through the many layers of sediment, working them as it cut, reworking them as it meandered. In depositing and redepositing alluvium, the river has both created and carved up its valley. Steep or sloping riverbanks exhibit walls of incised, older river material: ancient cobbles, lenses of crumbly sand and silt, angular debris tumbled down in steep, primeval sidestreams, and peaty remnants of ancient backwaters and marshes.

The geology of a river is often a buried world. To understand it, one must walk or float the channel looking for clues to its origins. The bedrock roots of mountains are rarely seen unless one goes sleuthing up the tributaries, which cut deep into rocky uplands (while the main river generally slices and reveals valley sediments). In the Sac's many feeder streams, a diverse bedrock record can be found: ancient limestones deposited in shallow oceans; layers of sandstone and mudstone subsequently transformed to metamorphic rock through the heat and pressure of mountain building; columns of cooled volcanic rocks that tower high above river valleys; marine sediments so rich in fossils they're known as *shell hashes.* Their names bear tribute to their *type locations* (where they are best and most characteristically exposed): Chico Formation, Pit Formation, McCloud Limestone, Castle Crags Pluton, Shasta Bally Batholith, Fort Jones Terrane. All are stony records of past eras during which seas advanced and retreated, volcanoes grew active and went dormant, mountains rose and wore down, millennia after millennia.

The Sac has taken its time in becoming itself. It has taken geologic time, as different from our daily sense of the clock as a moon shot is from our trips downtown. The past has an enormity our eyes can't see or our minds grasp. We use the words *geologic time* with a false familiarity; in truth, we can only find analogues for their meaning, as in: if all earthly time were condensed onto an annual calendar, the oldest exposed rocks on the planet date back as far as mid-March, plants and animals emerge in November, and humans don't show up until December 31. Somewhere in the vast span of that calendar, the river carved channels in arterial patterns that echo the natural branching of veins of leaves and vessels of animal blood. Over more time and into that calendar's month of December, rain-fed runoff coalesced into myriad channel paths that cut into the skin of soil. Once through the softer overburden, working into the calendar's final week, the river's already established braiding and meandering pattern was impinged on the magnificently diverse layers beneath: older sediments, ancient volcanic flows, and granitic bodies risen from deep in the earth.

Along the Sac, whether in the hills or on the valley floor, runs the floodplain—the corridor of relatively smooth and level land bordering the river and prone to flooding when the river runs high. Once drawn on early explorers' maps as a series of connected marshes that filled the Central Valley, the Sac's floodplain is now a narrow ribbon constrained by confining levees and adjacent, constrictive land uses. The floodplain possesses no single defining characteristic, but many. There is the *river channel* itself, steep banked or shallow banked and varying in width. There are channel *meanders,* bends in the river that in places ultimately evolve to be so long and looping that they circle back on themselves; eventually cut off, the isolated bends become smile-shaped *oxbows,* left independent of the channel. There are *point bars,* the beaches of sediment deposited on the insides of bends, often called sandbars and particularly attractive to summer swimmers. There are *meander scrolls,* the scars of old river bends, marked by strips of cottonwood seedlings that parallel the curving shore like parentheses. There are *sloughs*—areas of dead water—where isolated overbank flows linger, and *natural levees,* where floodwaters that overtop the channel banks build up berms or crests of deposited silt, sand, and clay.

Features of the watershed and floodplain express the native character of the Sac: its curling meanders; its abandoned sloughs and annually renewed forests; its cobble bars and mounds of natural levee; its willowy scent, its rushing sound, its wild floods, its quiet spells.

It's a living river, a world of dynamic shifts and pulsing ecosystems. It's no more a static web of ditches than the human body is a uniform, unchanging container of blood.

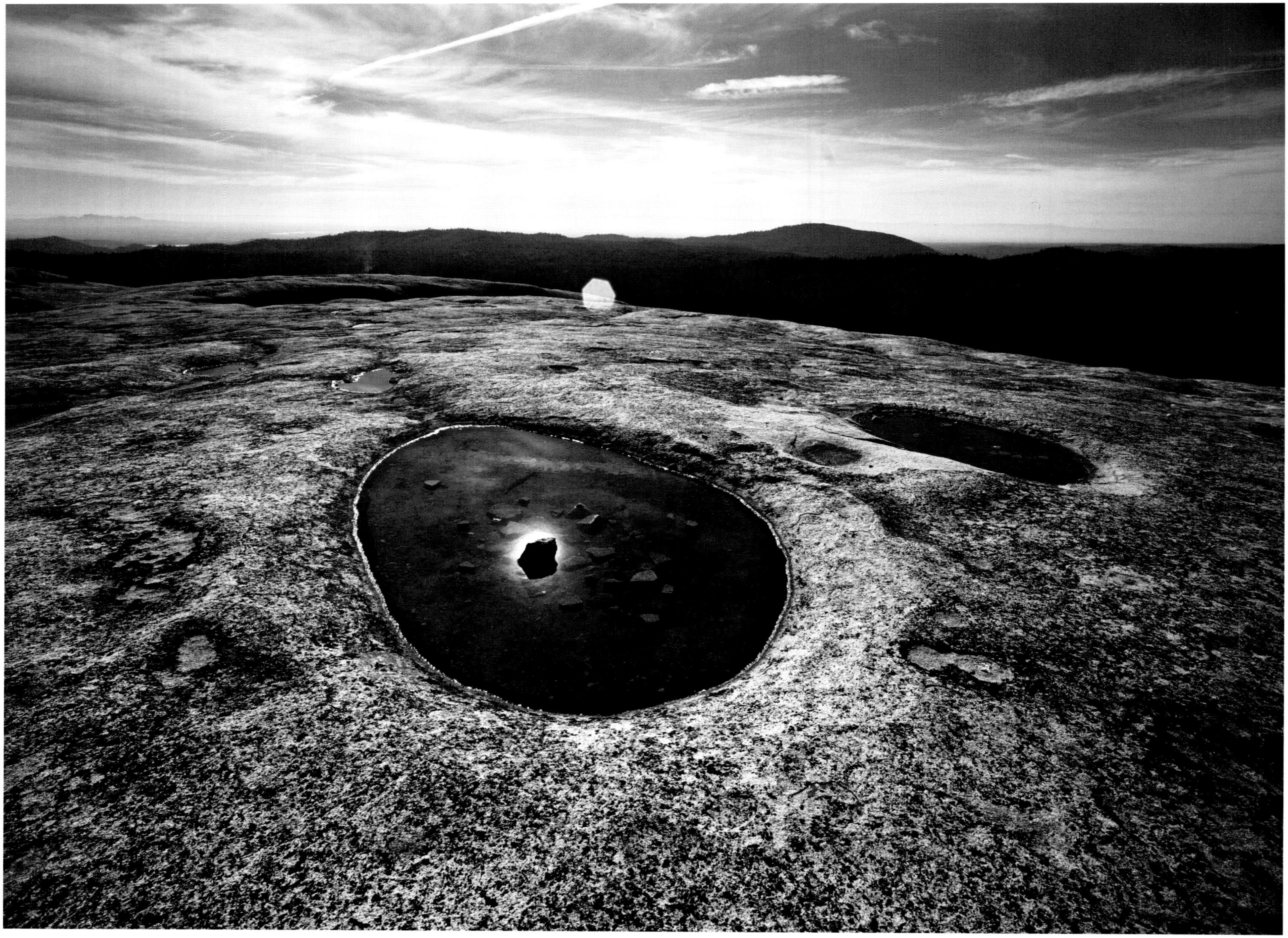

Last of the snowmelt on Bald Rock, Oroville, 2011

Aerial view of meandering Sacramento River, 1997 »

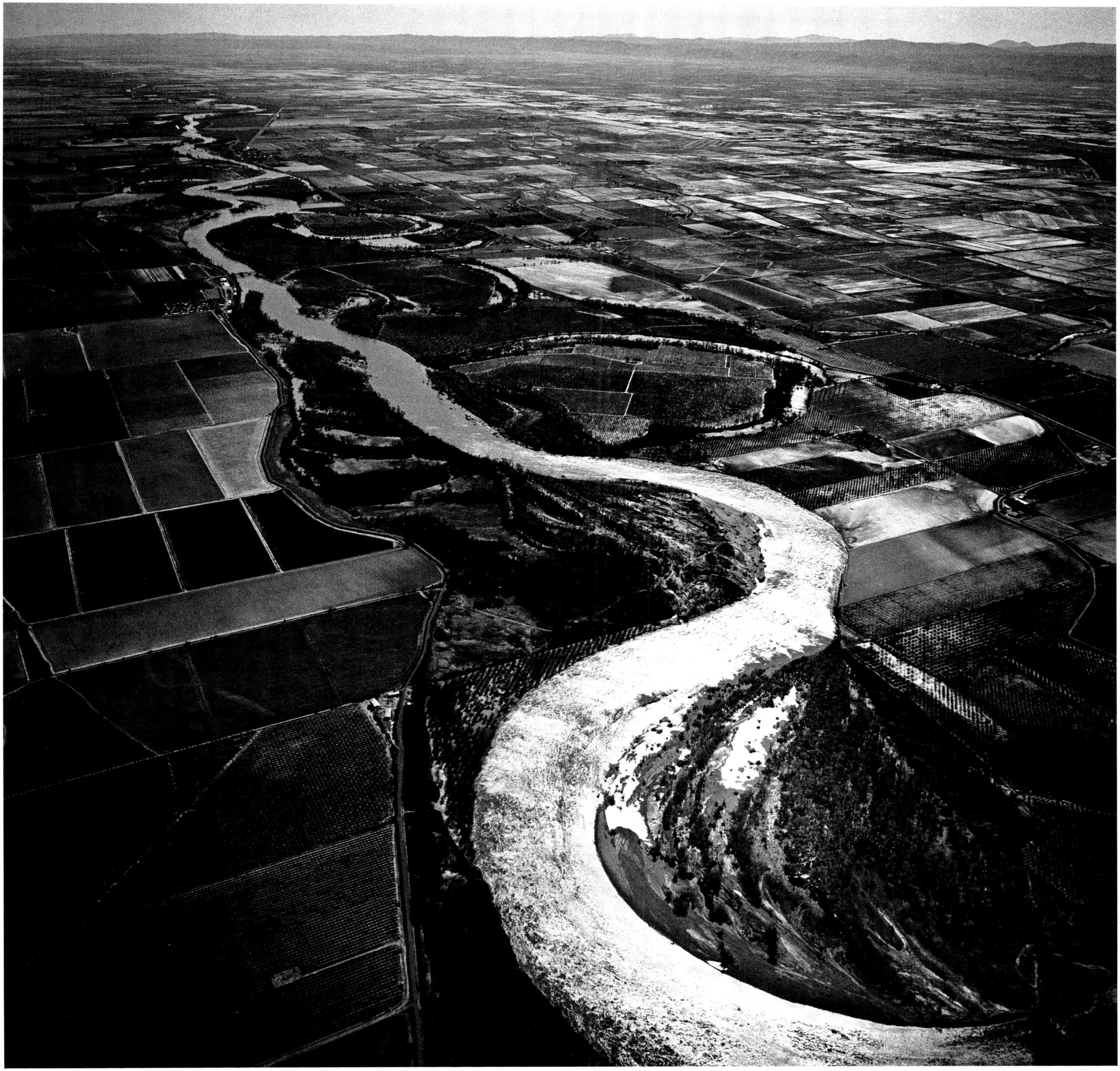

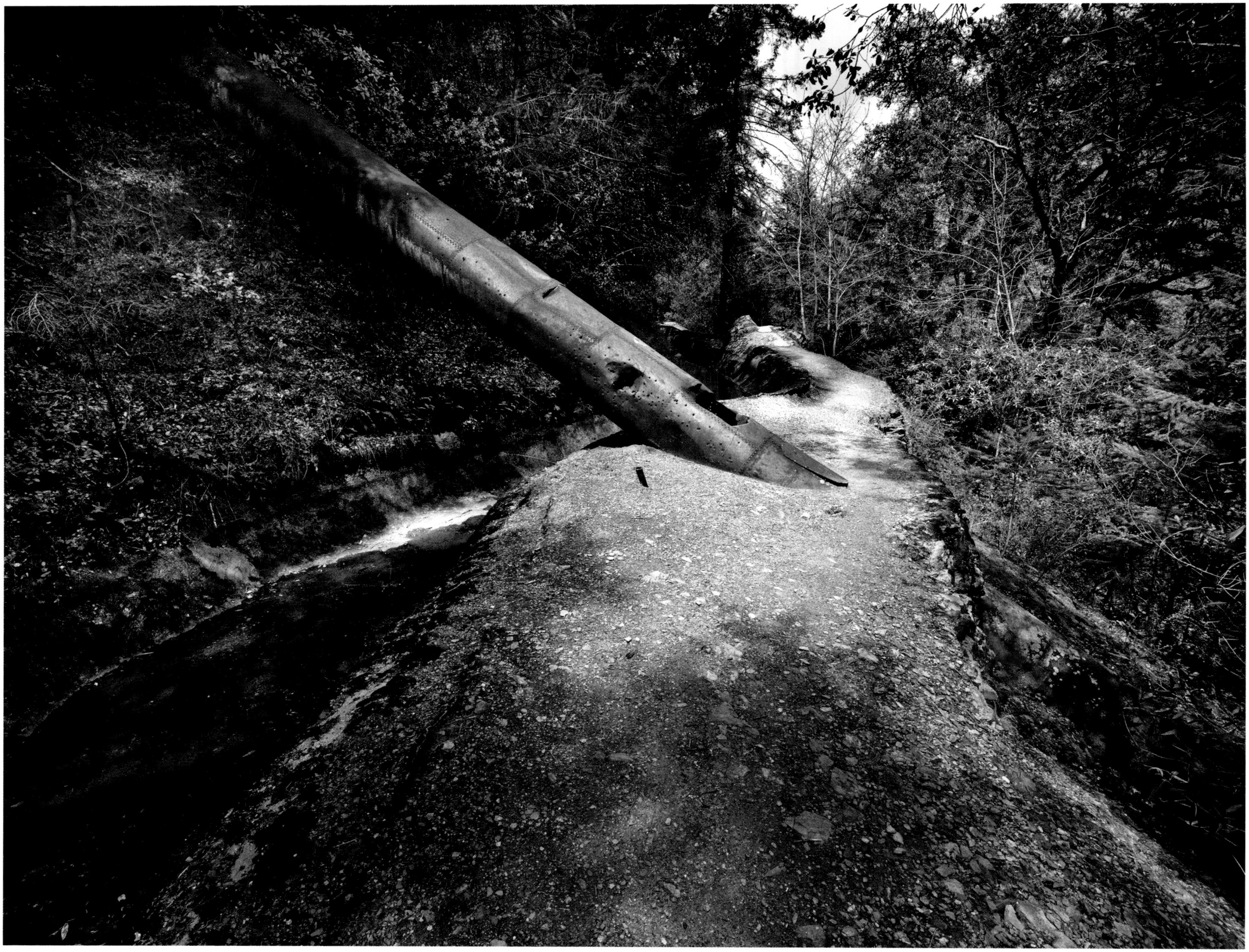

Where the Cherokee Siphon crosses the Miocene Ditch (in the 1870s, this pipe was part of a flume system that transported water to the Cherokee Hydraulic Mine on Table Mountain), Paradise, 2012

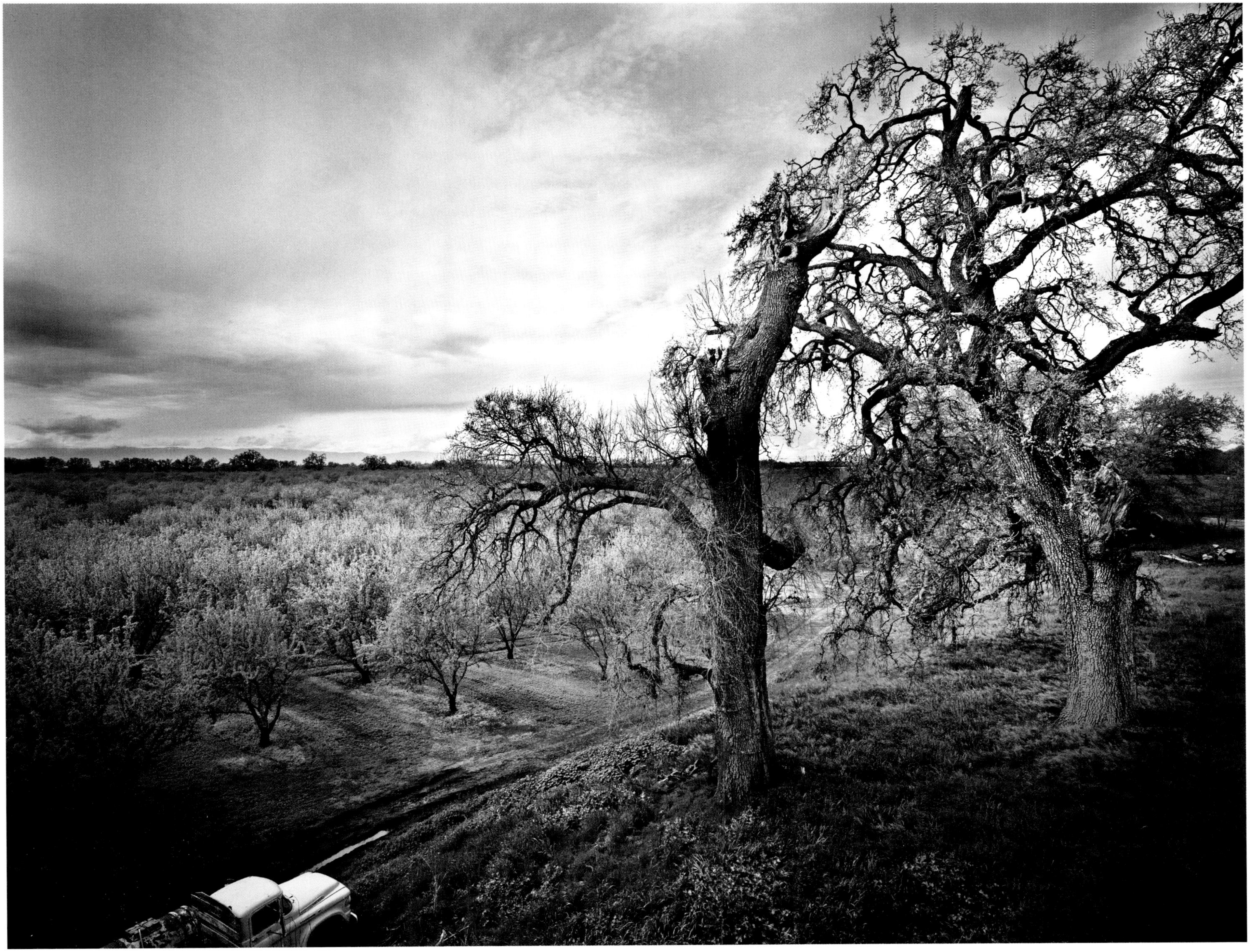

Oaks growing on the Modesto Formation just south of Red Bluff, 1999

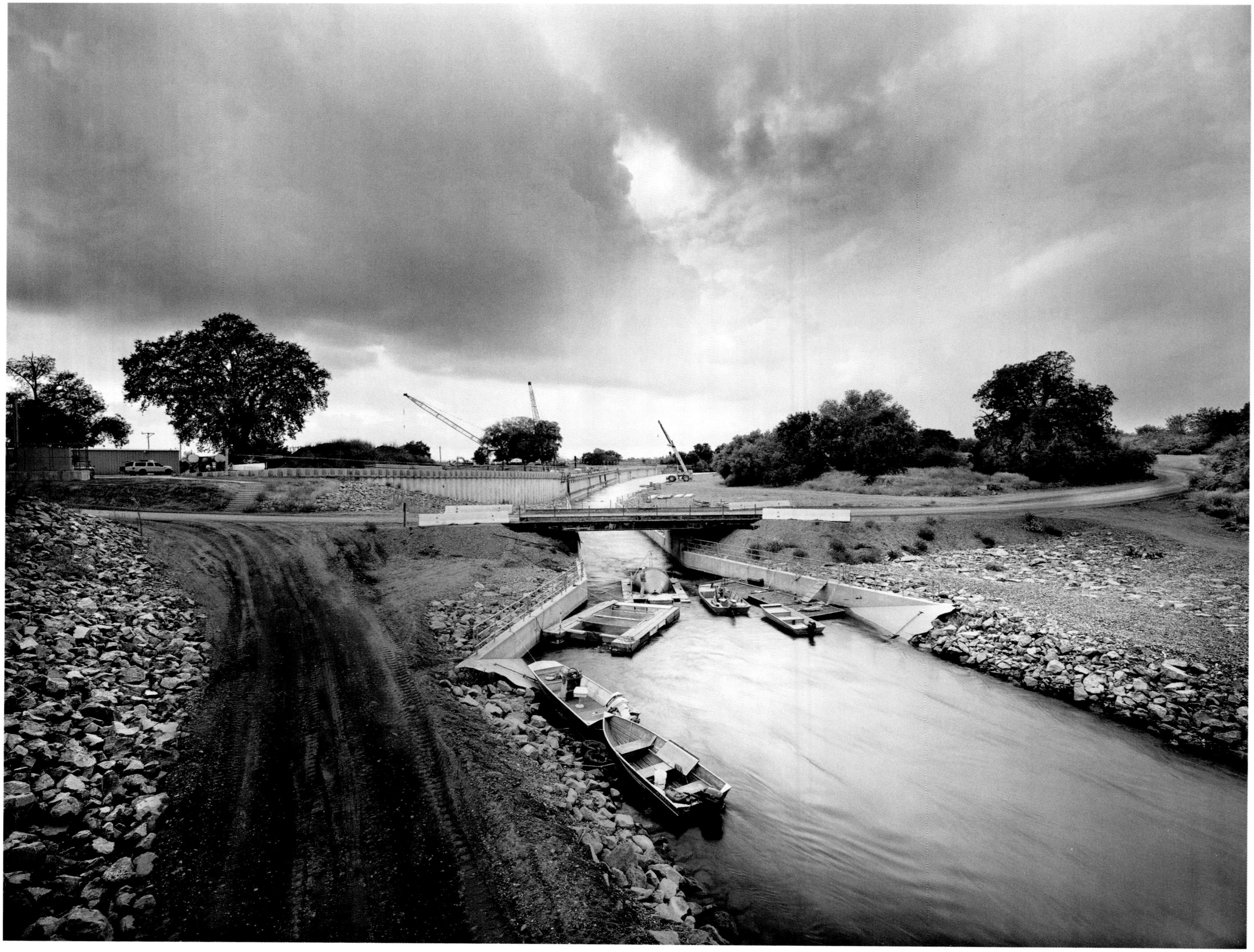

Glenn-Colusa Irrigation District fish screen, 1999

Geologist extracting a core sample at Parrott-Phelan Diversion Dam on Butte Creek, prior to reconstruction of fish ladder and installation of fish screen, 1995 »

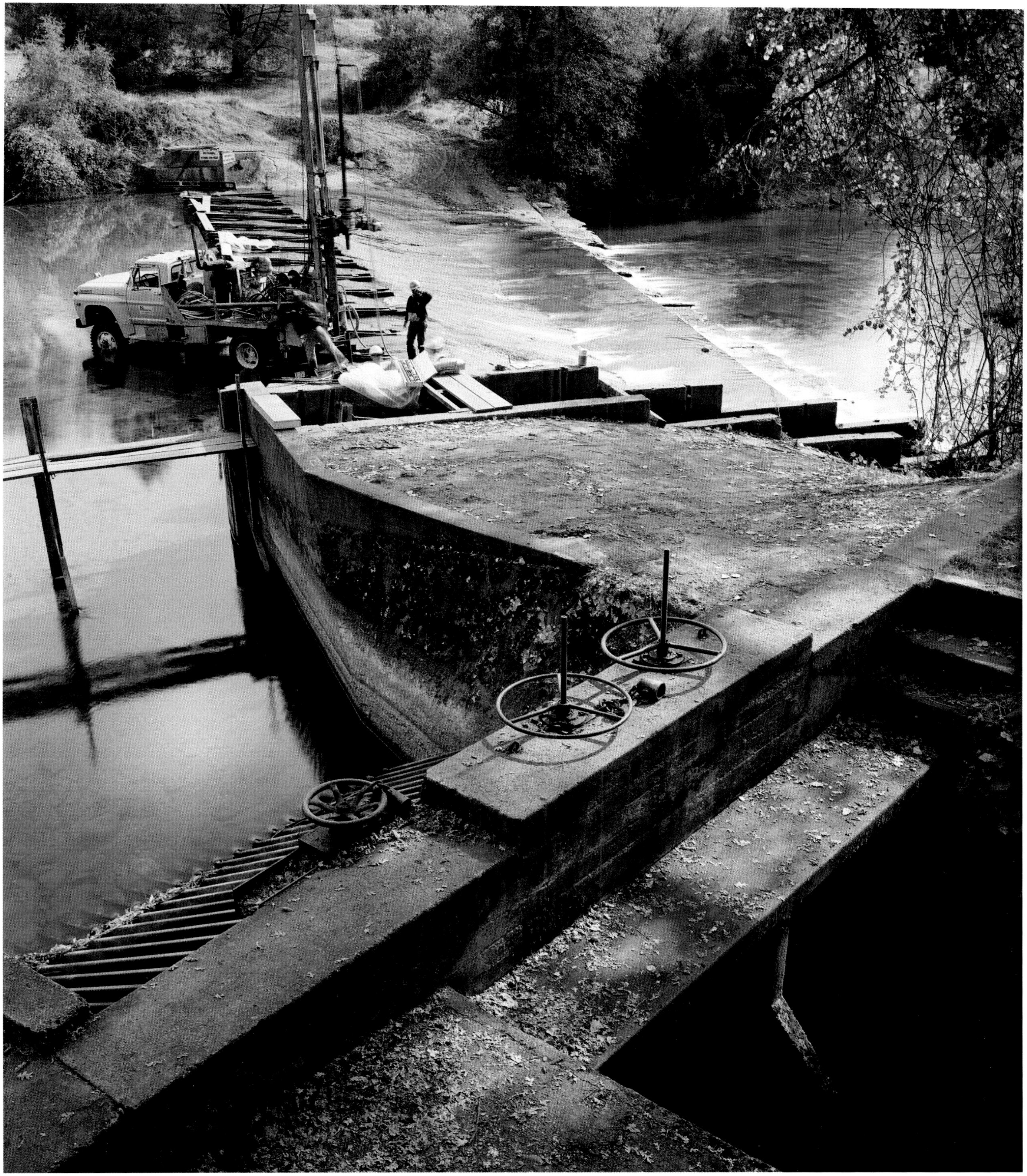

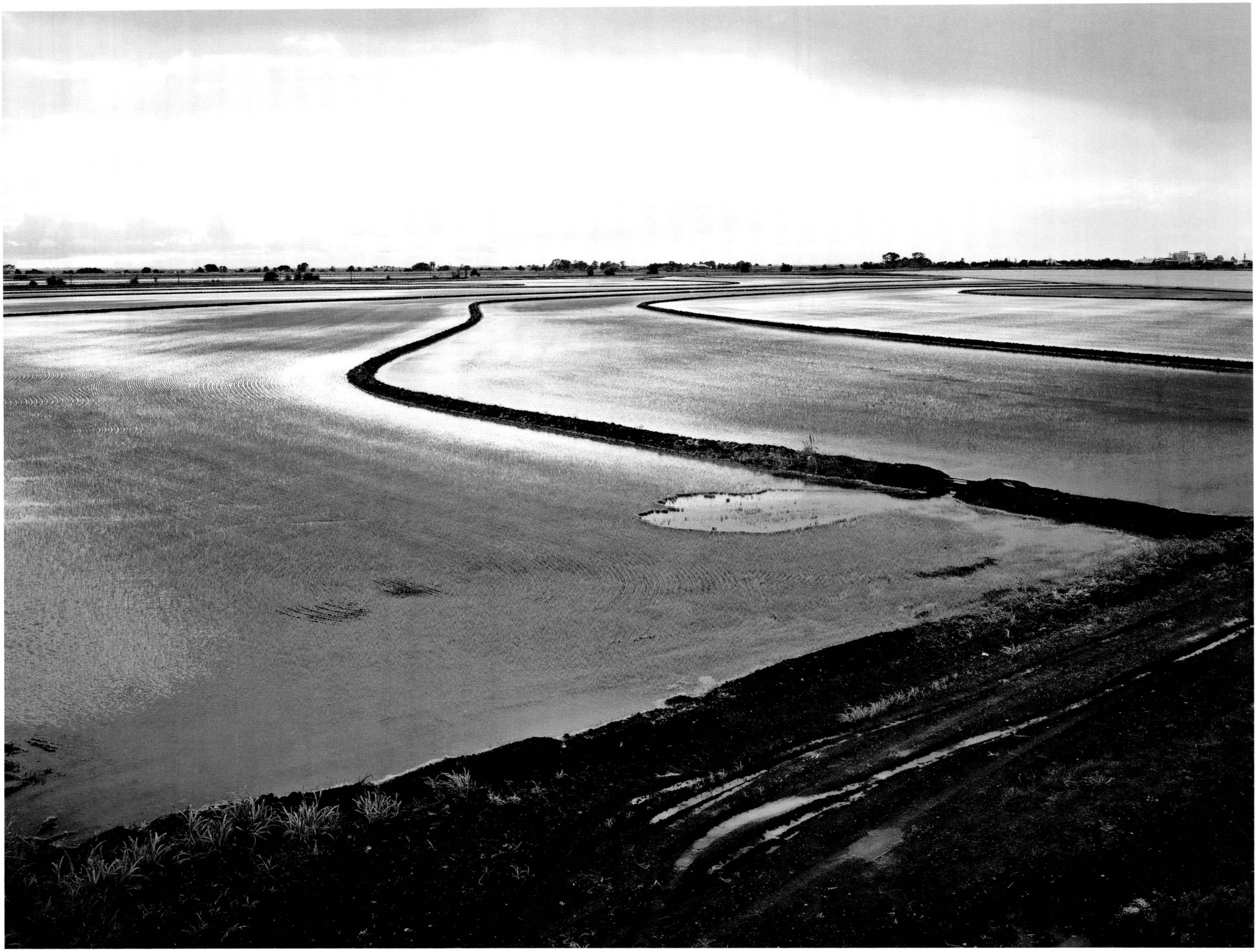

Flooded rice field near Richvale, 1997

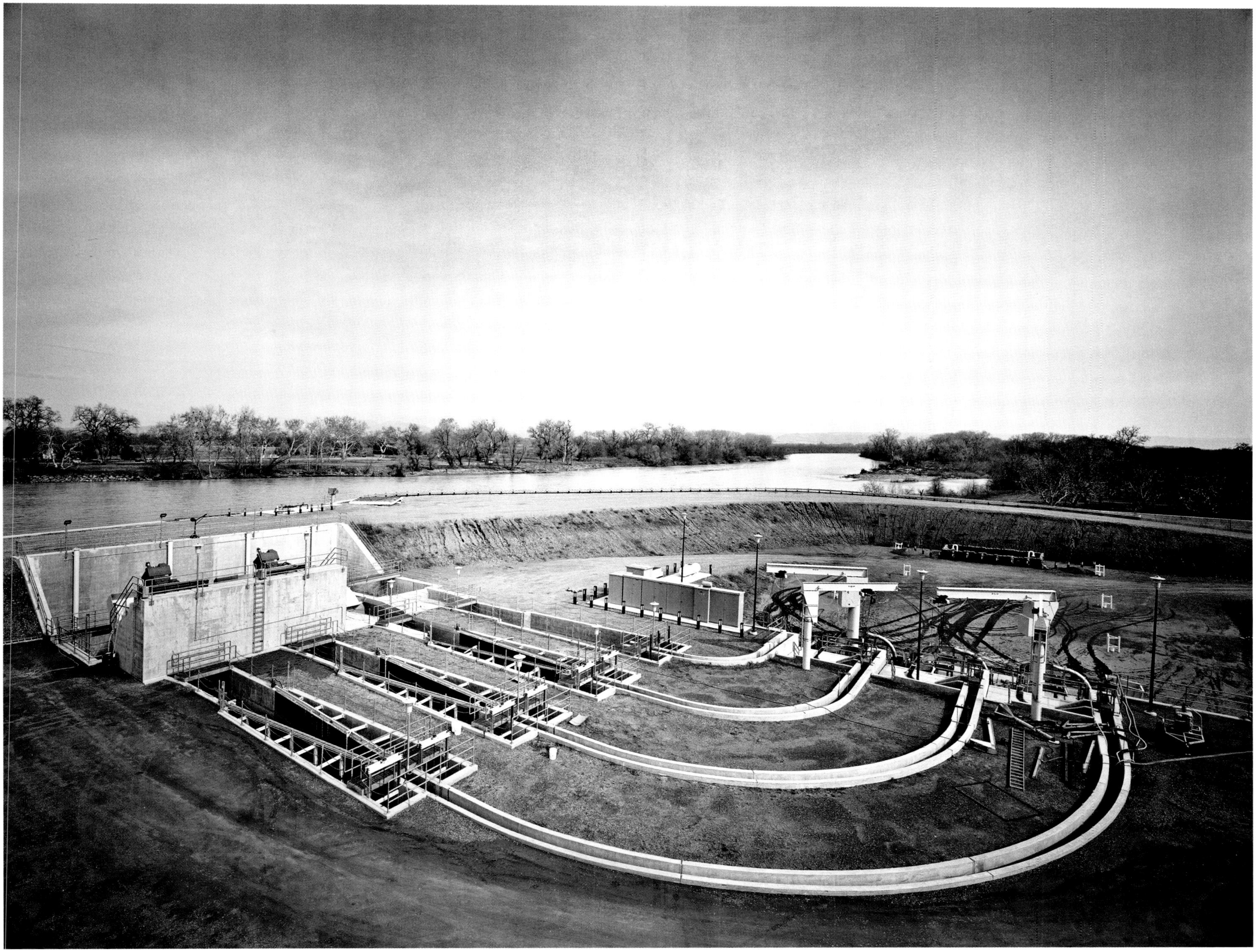

Pump and canal system no longer in use, adjacent to the obsolete Red Bluff Diversion Dam, Red Bluff, 1996

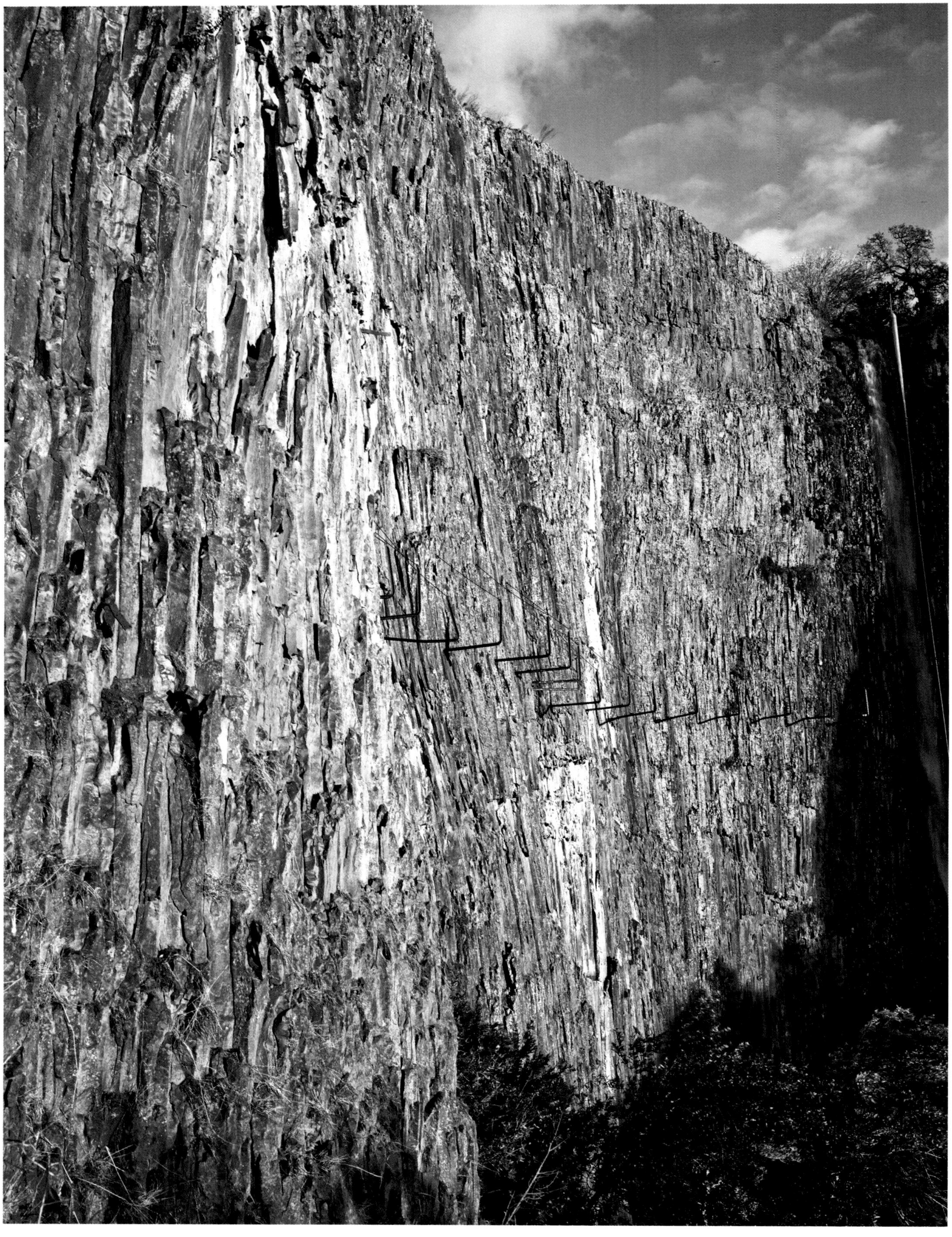

132-year-old brackets that used to harness a wooden flume on the basaltic cliffs of Table Mountain, 2012

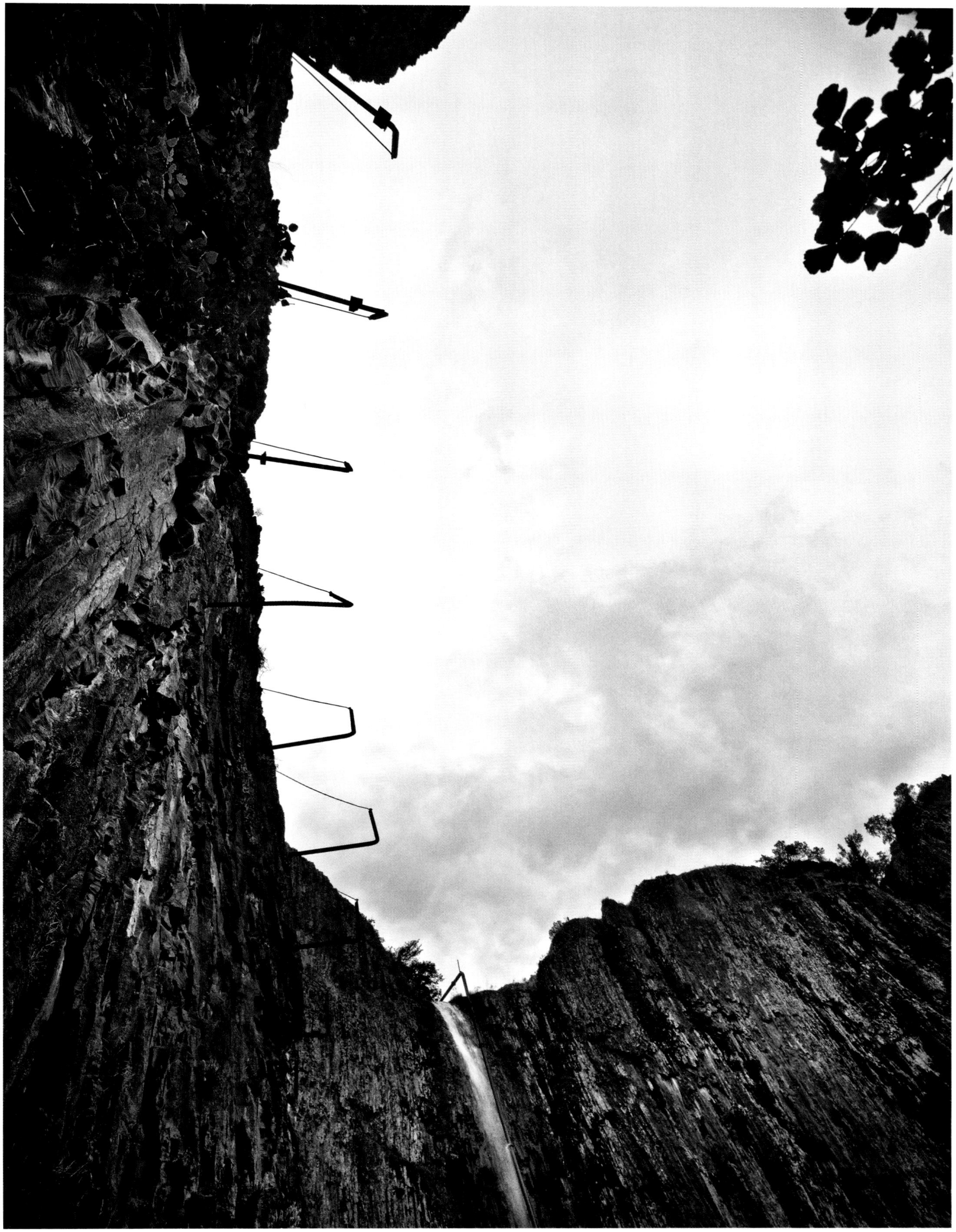

Brackets silhouetted against the sky, Table Mountain, 2012

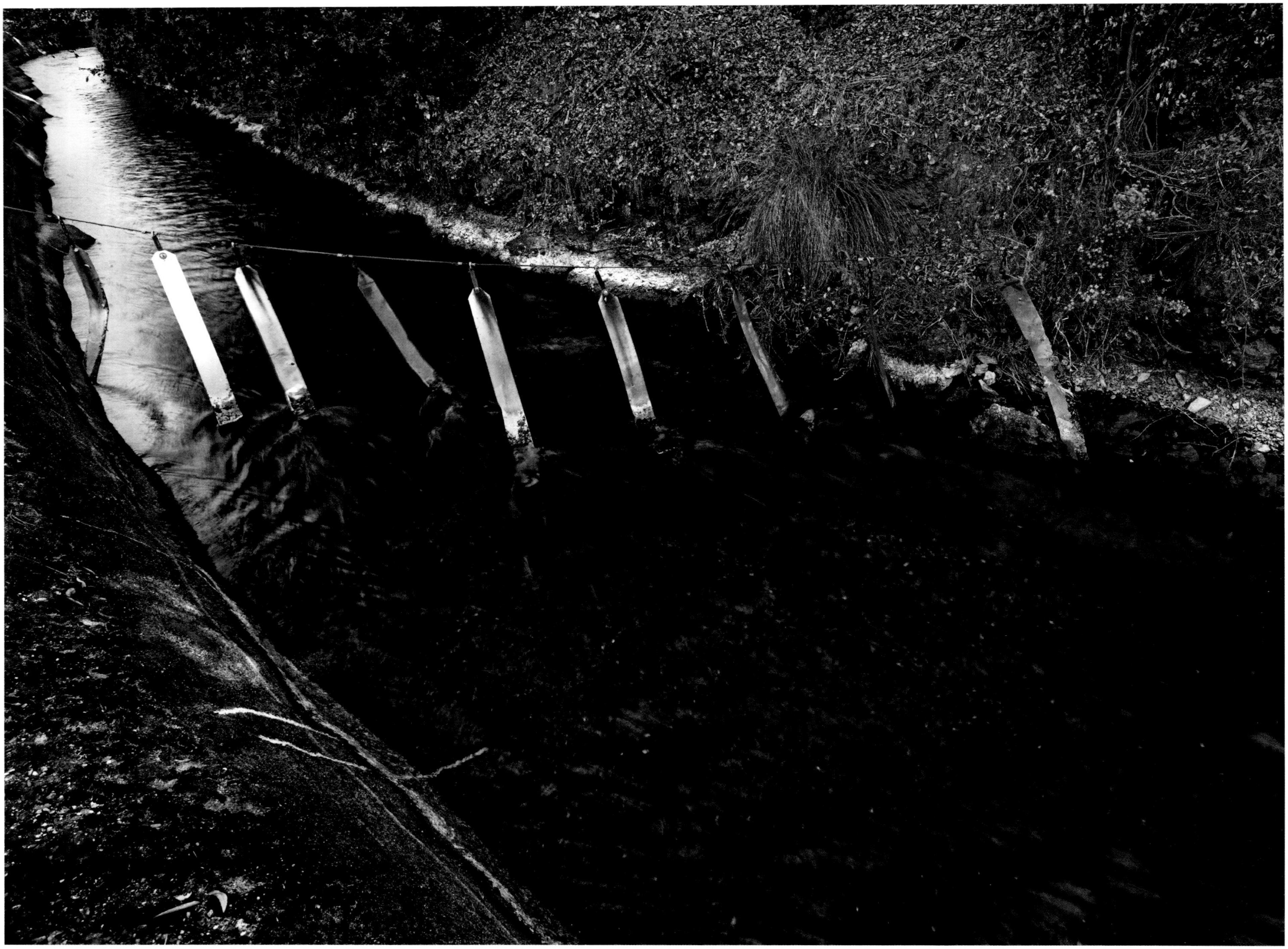

Deer guard on Centerville Canal, 2012

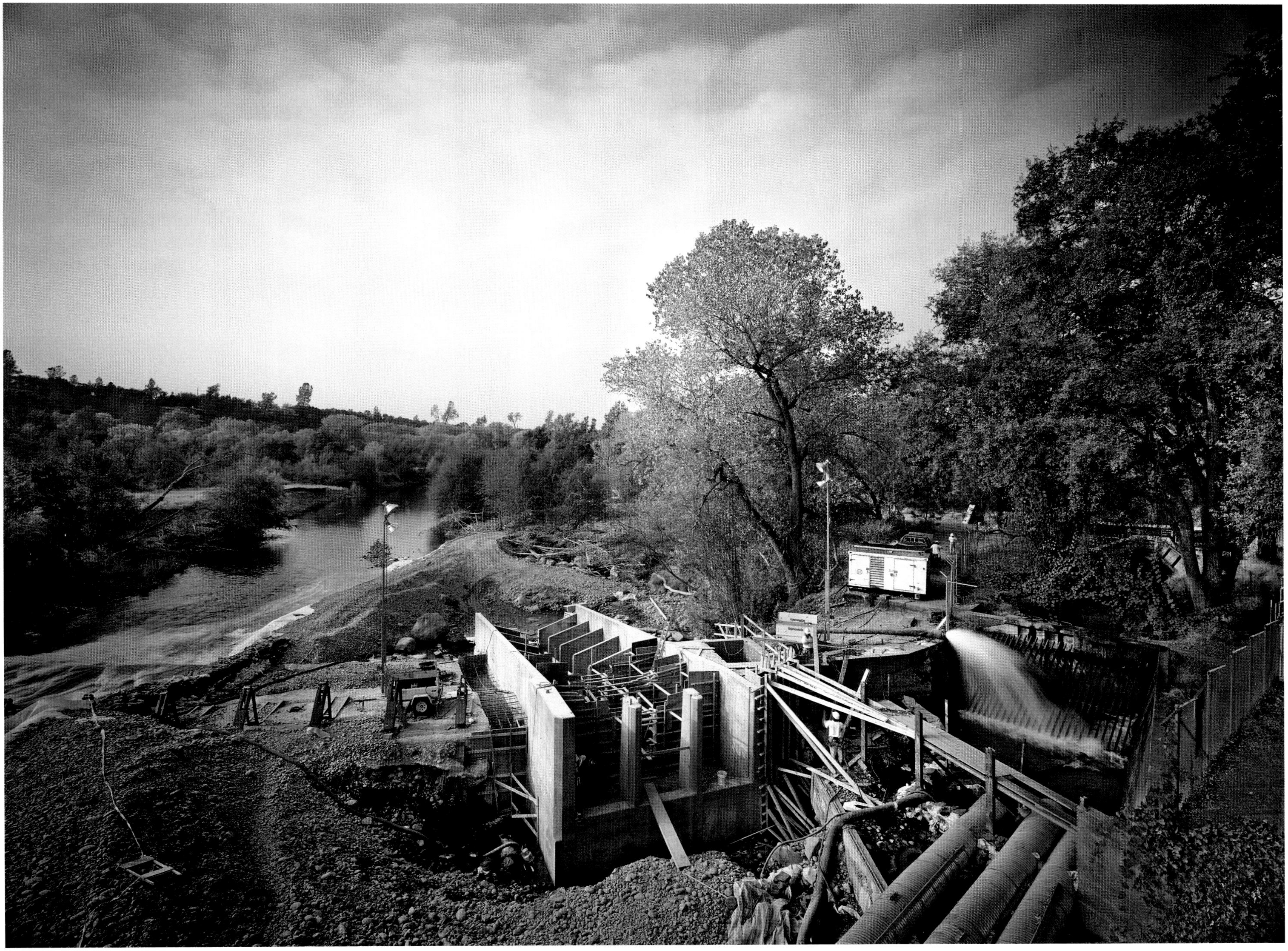

Parrott-Phelan Diversion Dam and Fish Ladder construction on Butte Creek, 1996

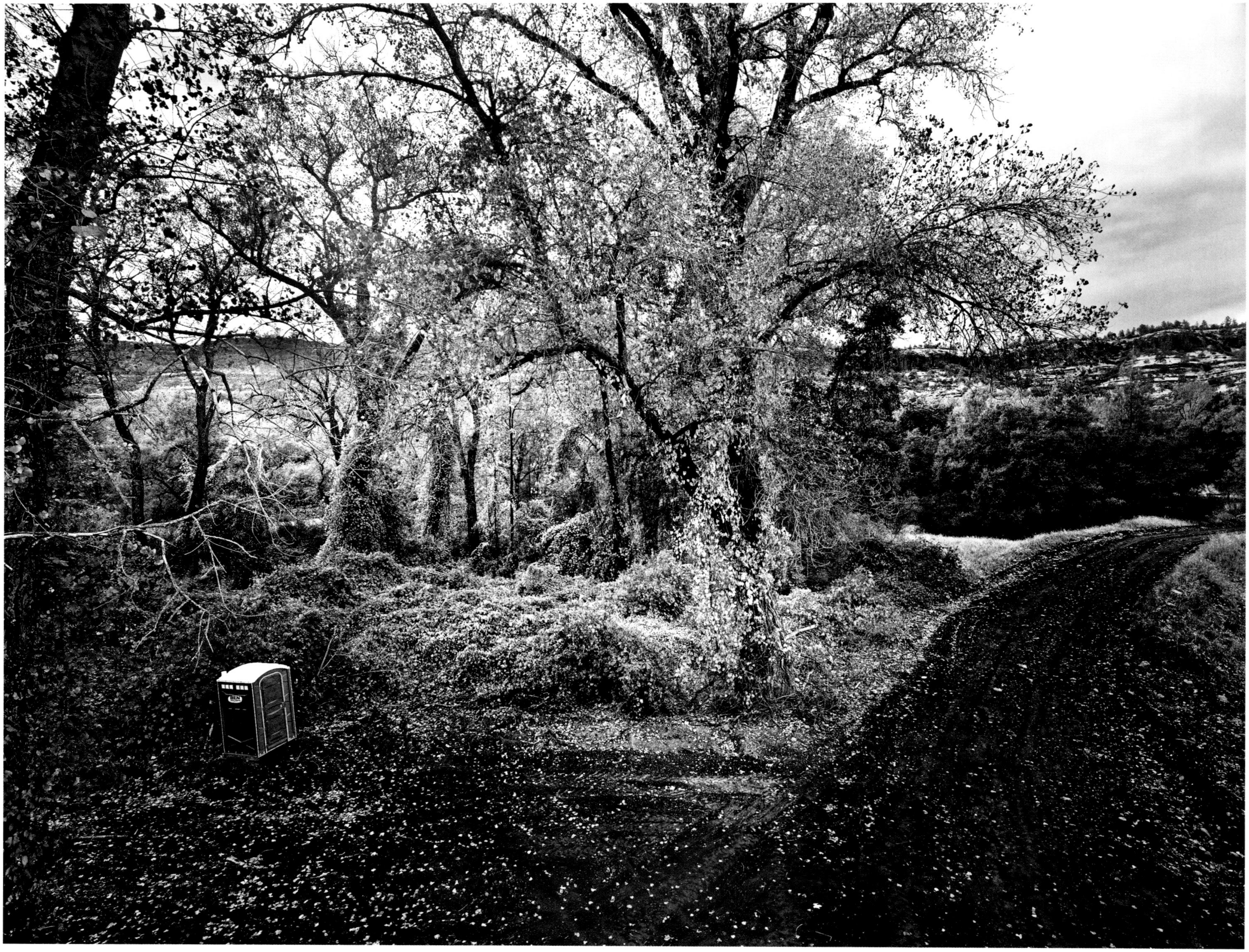

Riparian forest at flood-control site on Butte Creek, 1998

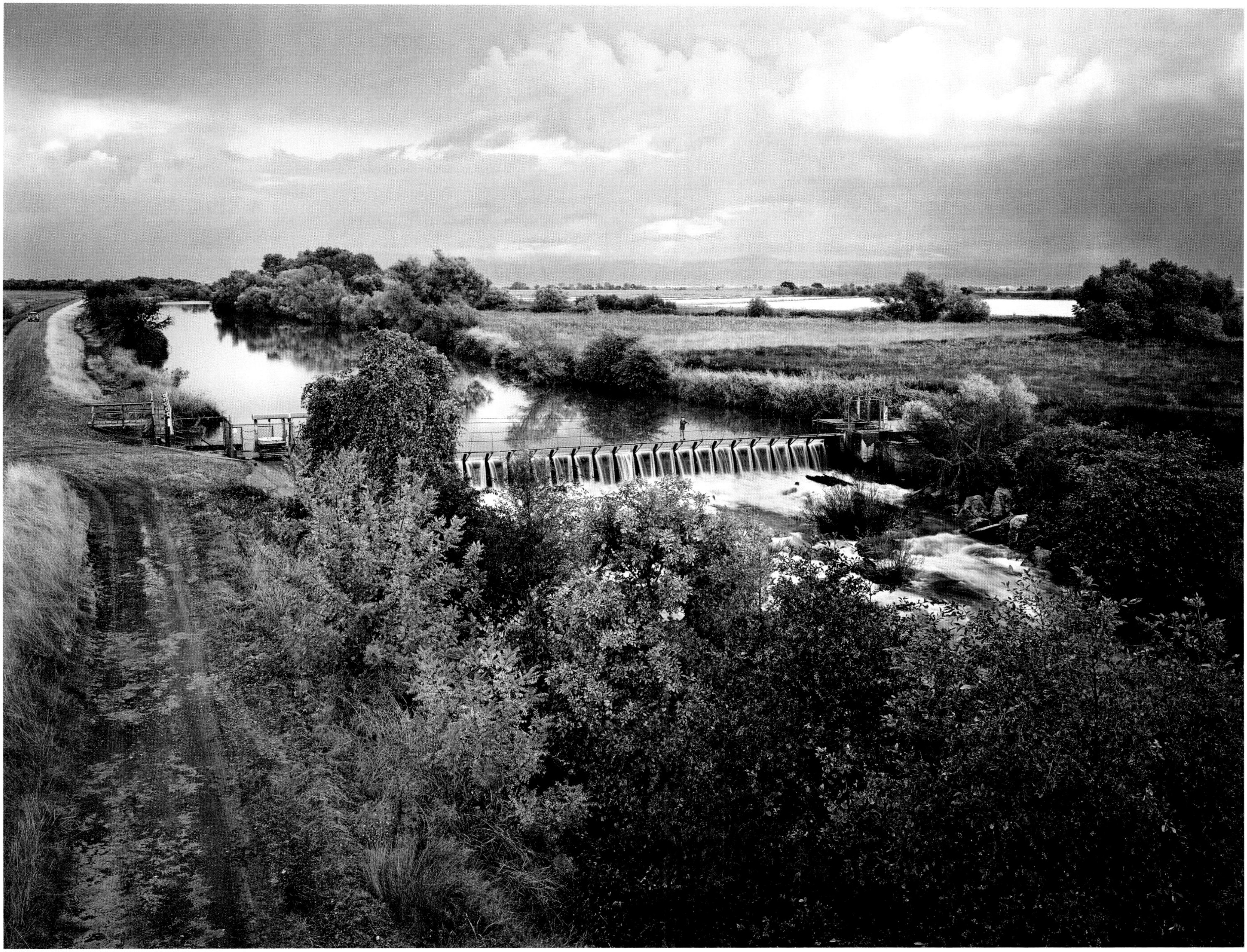

Western Canal Dam (this was removed and replaced by a subterranean siphon in 1999), 1995

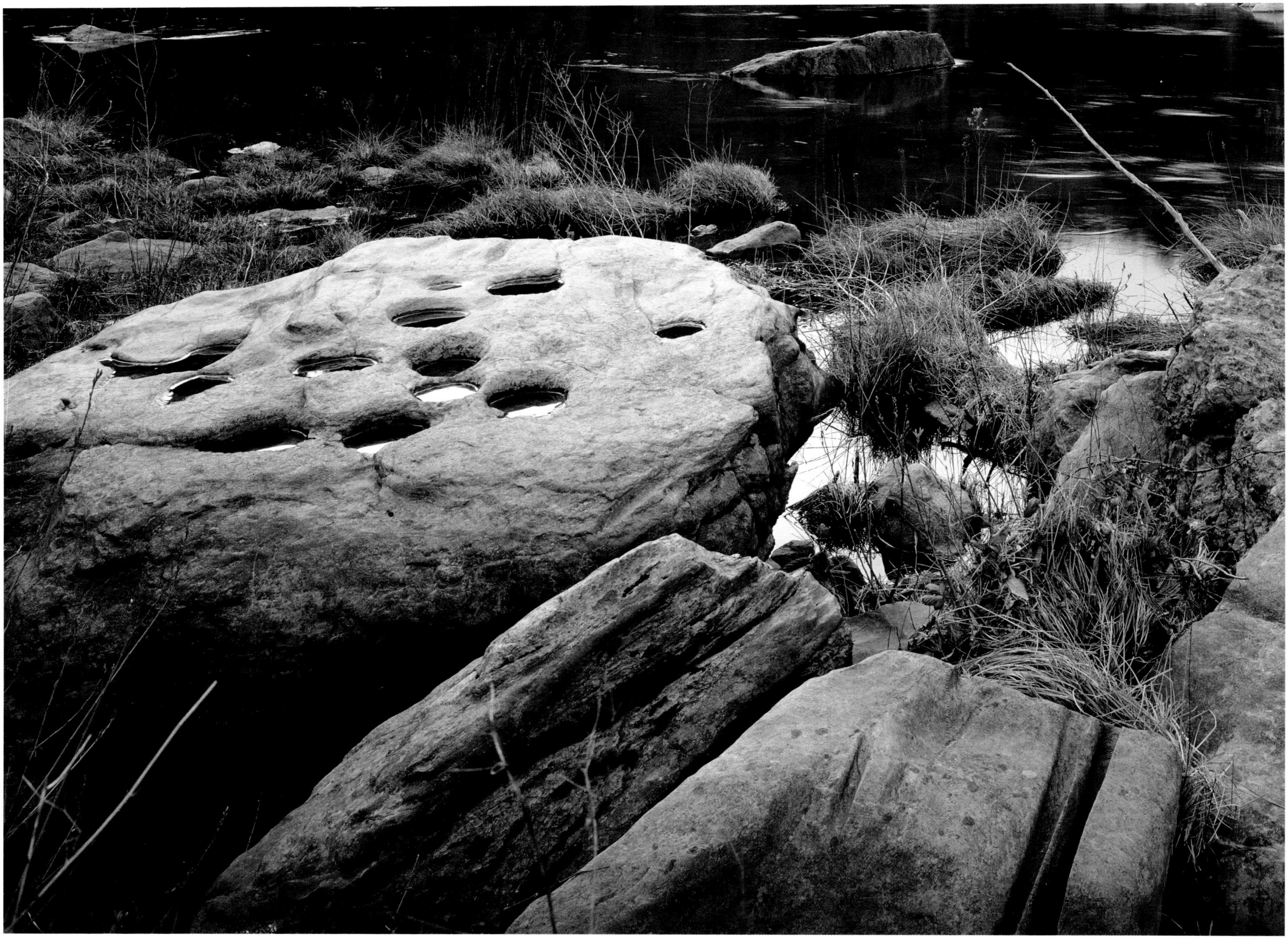

Two cultures in stone (Native American mortar holes and cable grooves from mining operations), 2002

I M M I G R A N T S

WE KNOW A RIVER DOESN'T SETTLE: it shape-shifts through the water cycle; it flows a crooked path; it is ever pushing for the sea. Nor did the first people settle when they reached the Sacramento Valley around twelve thousand years ago. They discovered a sustaining, watery world where marshes and rivers fused at the borders of grasslands and oak forests. Their lives cycled alongside the seasons of rivers and creeks. Following the migrations of wildlife when streams dried up in the valley, indigenous people made summer camp in the cooler, wetter high country. When winter rains swelled the river again, families left the shelter of the hills to return to low-lying marshes and tangled valley-floor forests.

As the first people immigrated to the Sac, a major extinction of the largest mammals in North America was under way: native cheetahs, mammoths, saber-toothed cats, short-faced bears, dire wolves, ground sloths, camels, horses, and peccaries were succumbing to the pressures of repeated glaciation during intermittent global ice ages. Decreased habitat and fewer food sources accompanied plunging temperatures. Most of the megafauna were gone from the Sac ecosystem by the nineteenth century. A fraction did survive: explorers' journals noted California grizzlies in herds grazing spring clover and wildflowers. Streams of deer, pronghorn, and elk flowed over the hills and throughout the valley. California condors, mountain lions, and coyotes roamed the savannah-like watershed; bison and wolves persisted long enough to be recorded in the first logbooks, but not in big numbers.

The habitat of the nineteenth century was still a naturalist's dream. Small and ephemeral tributary creeks and rivers to the Sac sustained streamside forests of oak, sycamore, ash, willow, walnut, cottonwood, and wild grape in swaths two-plus miles wide. Larger tributaries like the American boasted strips of streamside woodland up to four miles wide. The main-stem Sac was the crowned queen of river habitat, with a five-mile-wide streamside forest and nine hundred thousand acres of wetlands and rivers teeming with aquatic and land creatures. Geese and ducks blackened the sky, wings thundering.

Indigenous people lived in natural abundance for thousands of years. Clustered in family bands of twenty or thirty, the first humans adapted to the landscape, transforming it lightly, leaving little more than the traces of their fire stewardship in the burn scars of ancient trees. Over 120 different languages were spoken then in the geographic region that would become California: the richness of the culture and diversity mirrored the magnificence and variety of the land. The first nomadic humans established villages only around eight thousand years ago, as native populations in the Central Valley climbed to their peak of more than three hundred thousand.

Although they did not build canals, reservoirs, and cities, indigenous people did make their mark in stone and soil: fist-sized depressions ground with hand-milled acorns in the bedrock of riverbanks; foot paths that deepened game trails through grassland and forests; shells and shards discarded in midden piles in village sites; and bones buried in earthen graves.

The people passed down story, too, in a rich oral heritage: the watershed, they claimed, was tended wisely by the animal stewards and spirits who preceded them to the Sac. Every visible physical feature had been sculpted in some way by nonhuman forebears and natural phenomena, all with rich names and family associations inseparable from place. A prayer of the Yokuts people held, "My words are tied in one with the great mountains, with the great rocks, with the great trees, in one with my body and my heart." Every entity mattered; every soul was joined.

Spanish explorers first glimpsed the Sac from hills surrounding the delta in the late 1700s, and reported the presence of two great rivers and plentiful game at their terminus. Although the western edge of the continent was already somewhat settled because of the missions stringing up along the Coast Ranges, the need for excursions inland to find native labor to hire or enslave resulted in further exploration of the Central Valley. Thus Moraga encountered the Sac in 1808, fifteen miles north of modern-day Colusa on that holy day of sacrament. Navigations by ship followed, as well as fur trapping and additional forays inland on behalf of the missions. By 1823, European cartography showed the Sacramento Valley crudely but accurately ringed by unmapped mountain ranges and depicted as the first people had always known it: flat lying, with extensive, river-fed *ciénegas* (marshes). The water in the Sac watershed was not the arm of ocean that Cortez in the 1500s speculated might separate California from the continent, a concept promoted on maps for more than two hundred years. The valley brimmed with fresh water, in great quantities.

Today the water is largely confined to the flumes, canals, and ditches traversing the watershed for irrigation and urban water use. Transformation began on a small scale, just as the Sac's abundant riparian forest must have started with a single seed.

Spanish immigrants instituted irrigation for the missions, following European tradition based in Roman and Eastern European hydraulic engineering. The Spanish followed another tradition as well: limited use of water, as mandated by a strong central government accustomed to exercising authority for the good of an arid land culture. Small, well-planned systems of ditch and stream withdrawal didn't alter the face of the Sac watershed greatly. Only later did transformation occur on a landscape scale, by large water projects initially founded by private parties in the spirit of free enterprise.

When it came, extensive watershed change arrived with startling speed. After the discovery of gold at John Sutter's American River sawmill in 1848, the Sac's tributaries attracted immigrants in pursuit of quick riches. To sort "color" from quartz-rich streams, they first washed handfuls of gravel in slope-sided pans; later, for greater efficiency, they used high-pressure hoses to flush entire hillsides of gold-bearing alluvial gravels through sluice boxes. This ingenious but highly erosive *hydraulic mining* technique required water, and lots of it: within ten years of gold discovery, private businesses had developed seven hundred miles of canals and ditches for the hydraulic mines in Nevada County alone; by 1879, when hydraulic mining peaked as an industry, that same system had grown 50 percent larger. And, unlike the Spanish mission authority, no one was in charge, certainly not the government.

The growing waterworks spun a new web on the land. Immigrants filled blank spots on the map, founding towns, farms, and businesses to keep pace with the gold boom. They planted orchards, introduced cattle and dairy stock, plowed fields, diked and drained wetlands, cut oak and conifers, and grew hops, grapes, wheat, rice, and other highly prized crops. Towns bloomed like spring gardens and required a massive system of irrigation.

Indigenous people waned. They were decimated by fighting and disease, as well as moved by treaty and force to unfamiliar land in the Coast Ranges. By 1900, only twenty thousand indigenous people lived in what is now California, their cultures threatened and many languages lost. Anthropologists and historians rushed to catalogue their unraveling cultural systems, devising the tribal names we know today: Miwok, Yuki, Pomo, Yokuts, Maidu, Modoc, Wintu, Shasta, and more.

In this dark period of our history, loss piled onto loss. The Sac's streamside habitat was cut and bulldozed; today only 10 percent of the original riparian corridor remains. The California grizzly was tortured for sport and trapped for trophy; by 1908, it had been hunted to extinction. Deer, elk, and pronghorn were shot and fenced out of former habitat: they no longer stream in great herds over the hills. Hunting and loss of habitat exterminated a large part of the annual waterfowl migration as well; it is a fraction of the awe-inspiring spectacle it was at its peak.

The delta was cultivated from over 1,100 square miles of coastal brackish marsh and estuary to a patchwork of farmland seamed by levees and roads. Potatoes, sugar beets, asparagus, pears, and other crops sprung from rich peat soil built of fine sediment and organic matter brought by the rivers. California's population doubled: the 1868 census recorded 380,000 individuals, not counting the state's native population, which had recently reached nearly that number. Less than a century following European contact, a new day had come to the Sac. Immigrants had arrived not only from northern Europe and the United Kingdom, constituting the greatest number of California's 1800s population, but also from Asia, South America, and Africa.

Water in creeks and rivers had by tacit agreement become a commodity—one that could be moved by anyone; one that if so moved could profit entrepreneurs immensely. More people would come, millions of them, who would view taming the river as their birthright: if not the whole river, certainly the water within it. The Sac was there not to sustain the diverse life within and along its shores, but for our use.

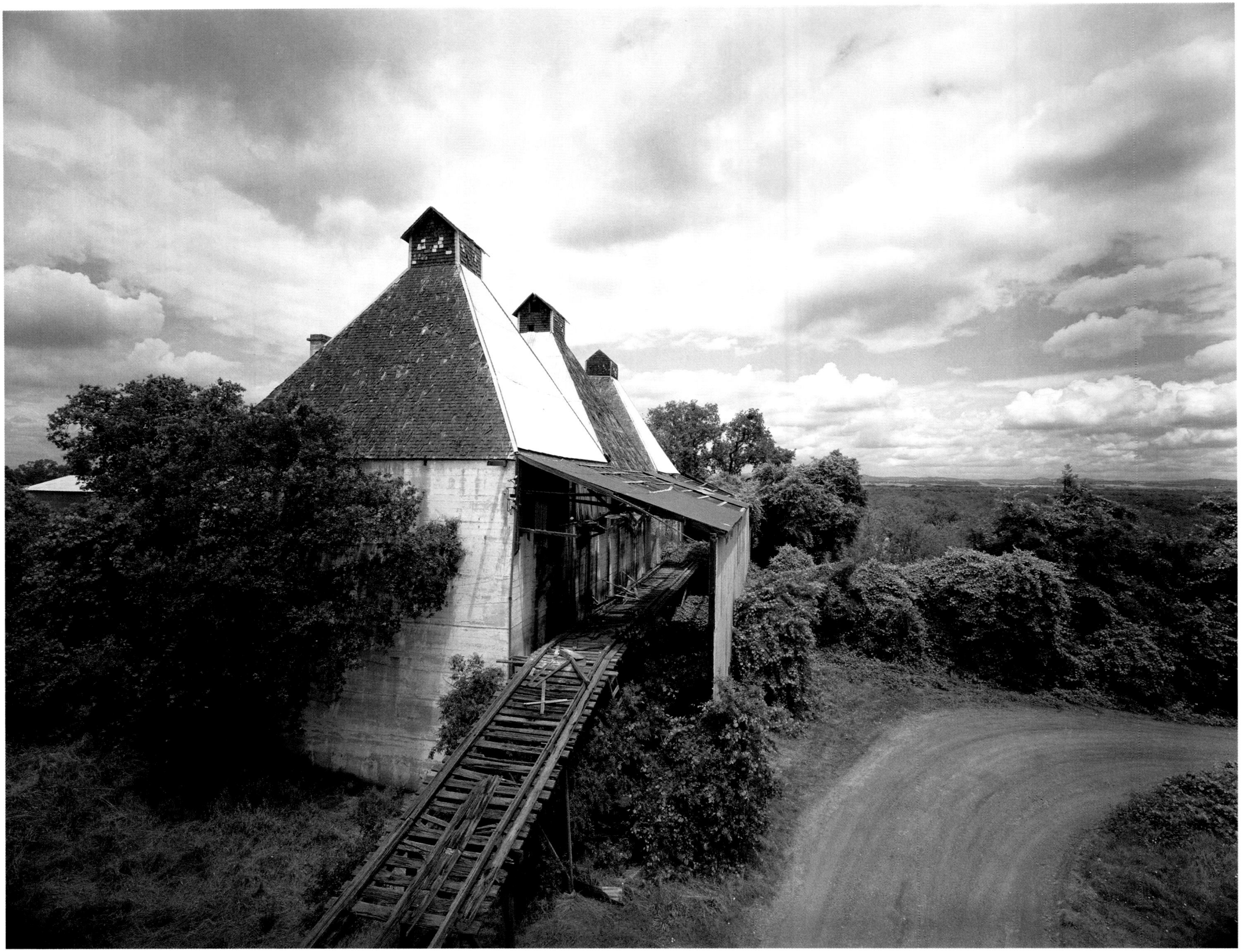

Buildings for drying hops, which grow here as a result of soils made rich by alluvial deposits from upstream hydraulic-mining sites, Wheatland, 1999

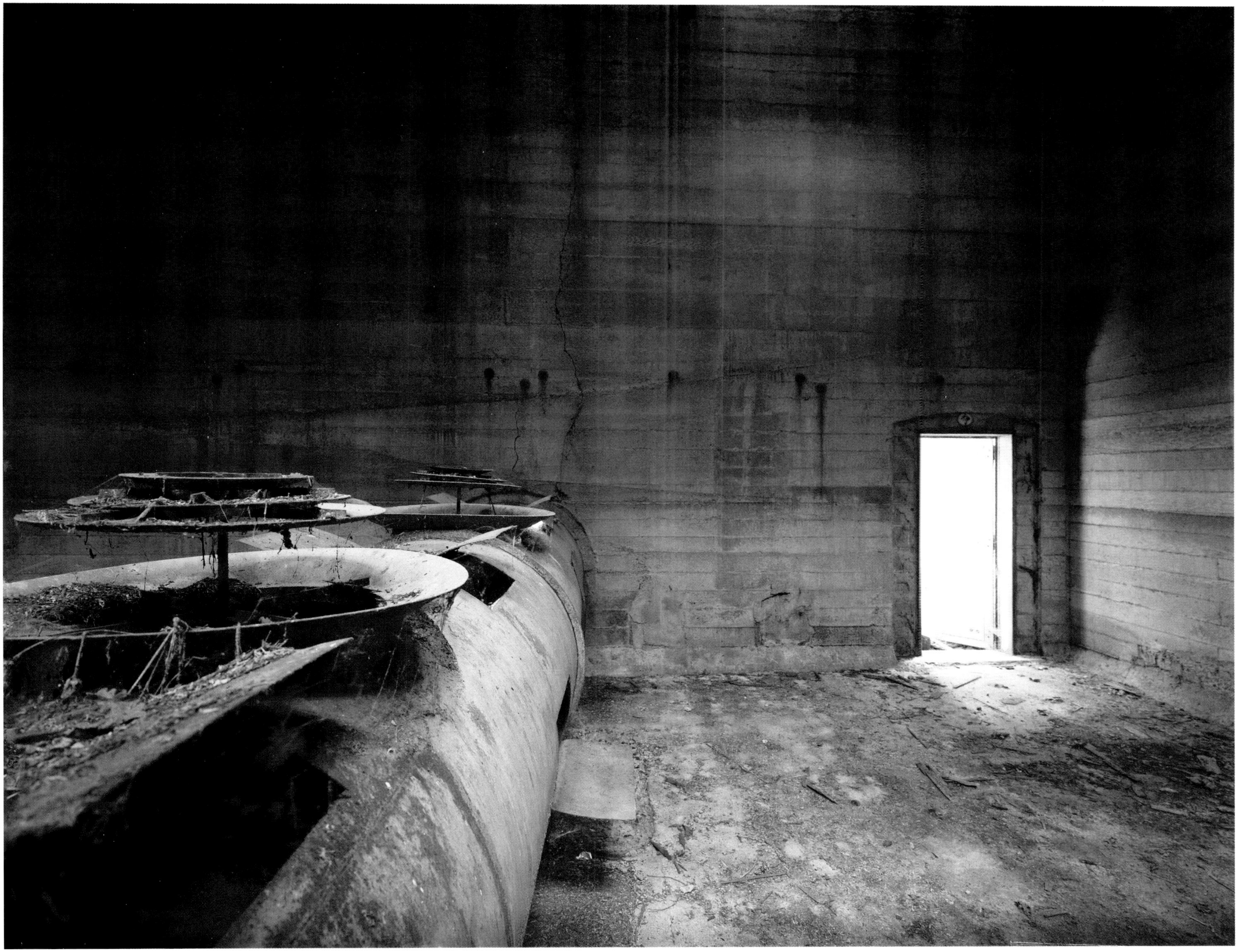

Interior of hop-dryer building, 1999

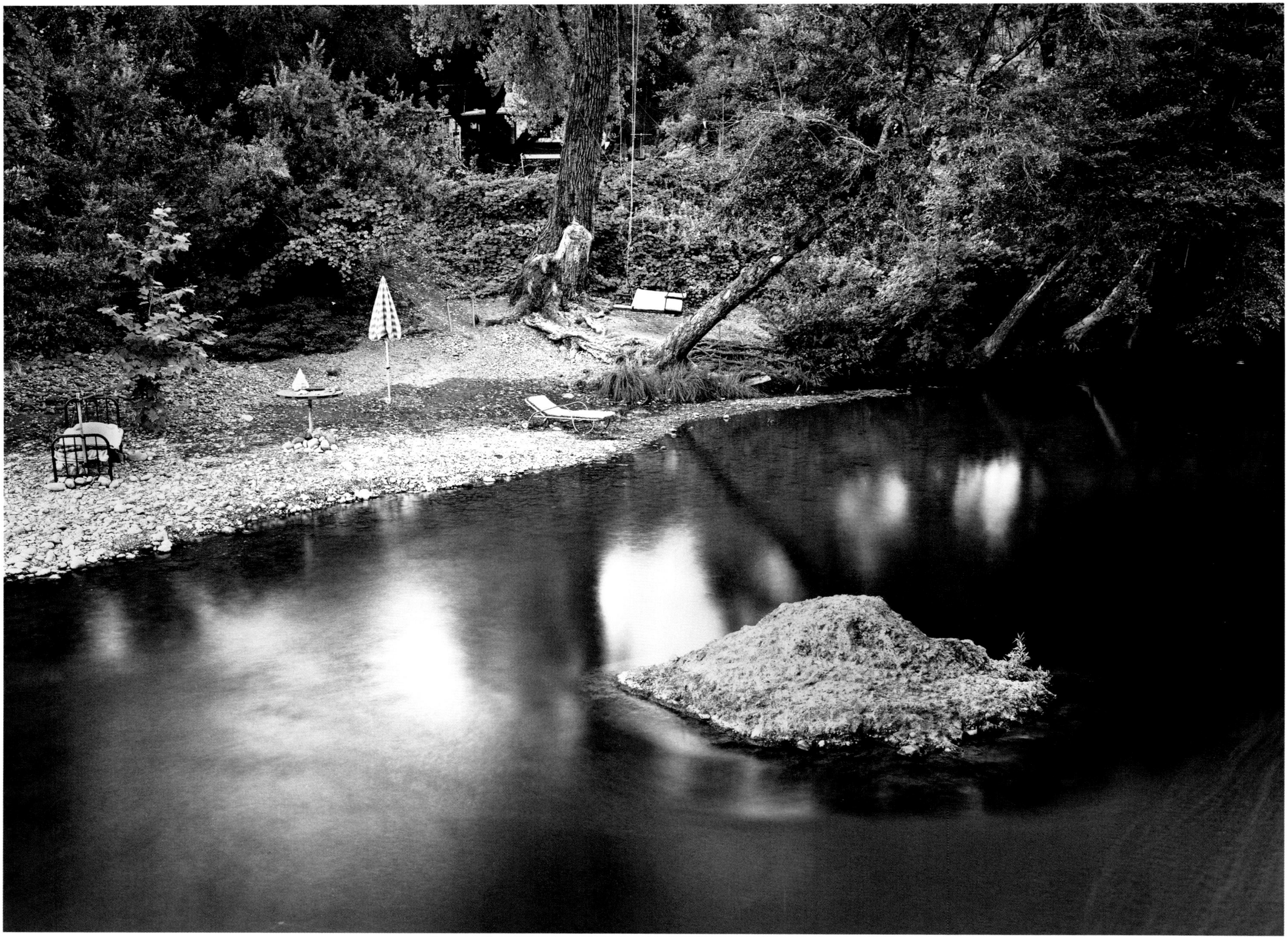

Lloyd's place on Butte Creek, 1997

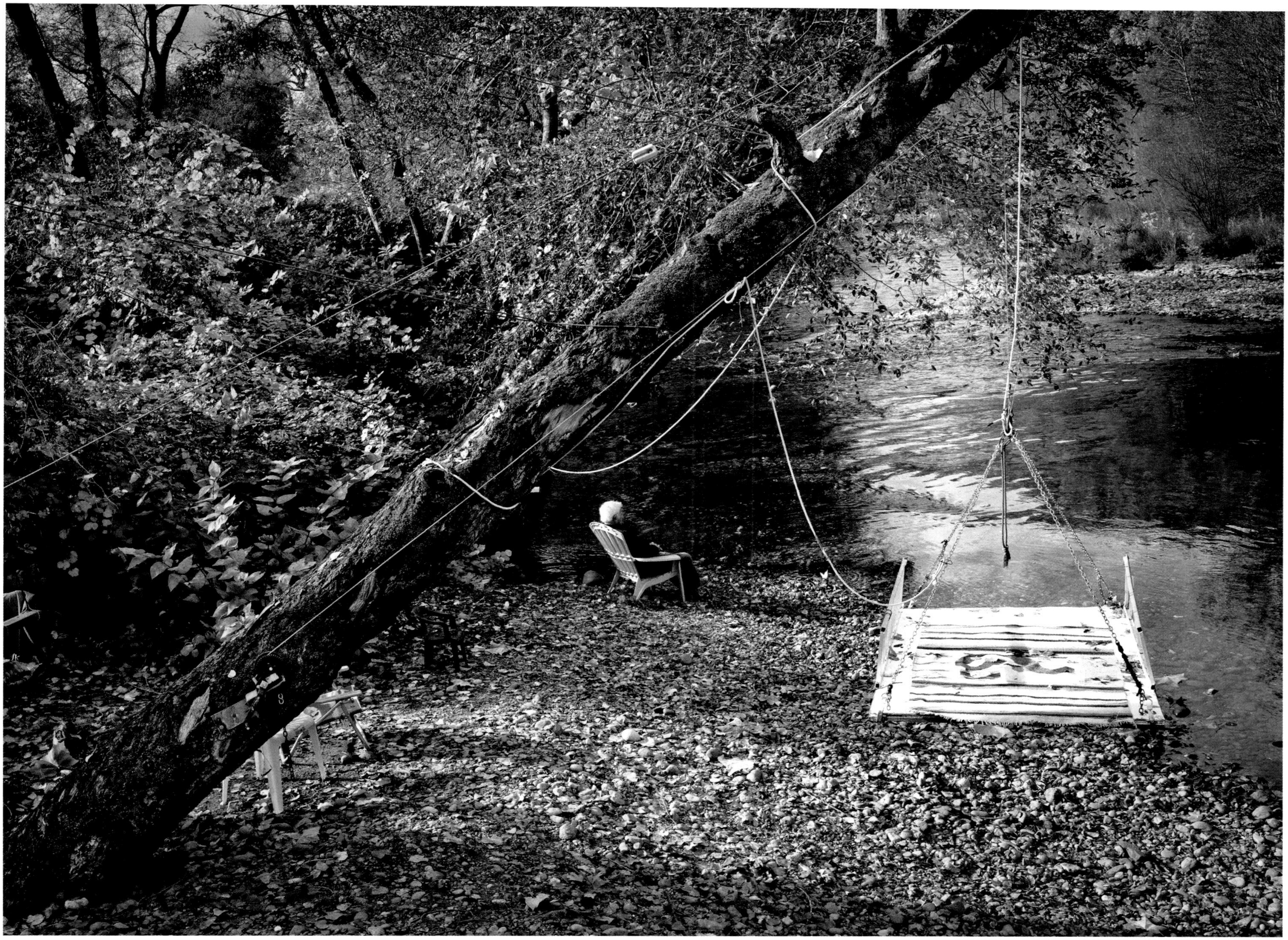

Lloyd's place, Butte Creek, 2012

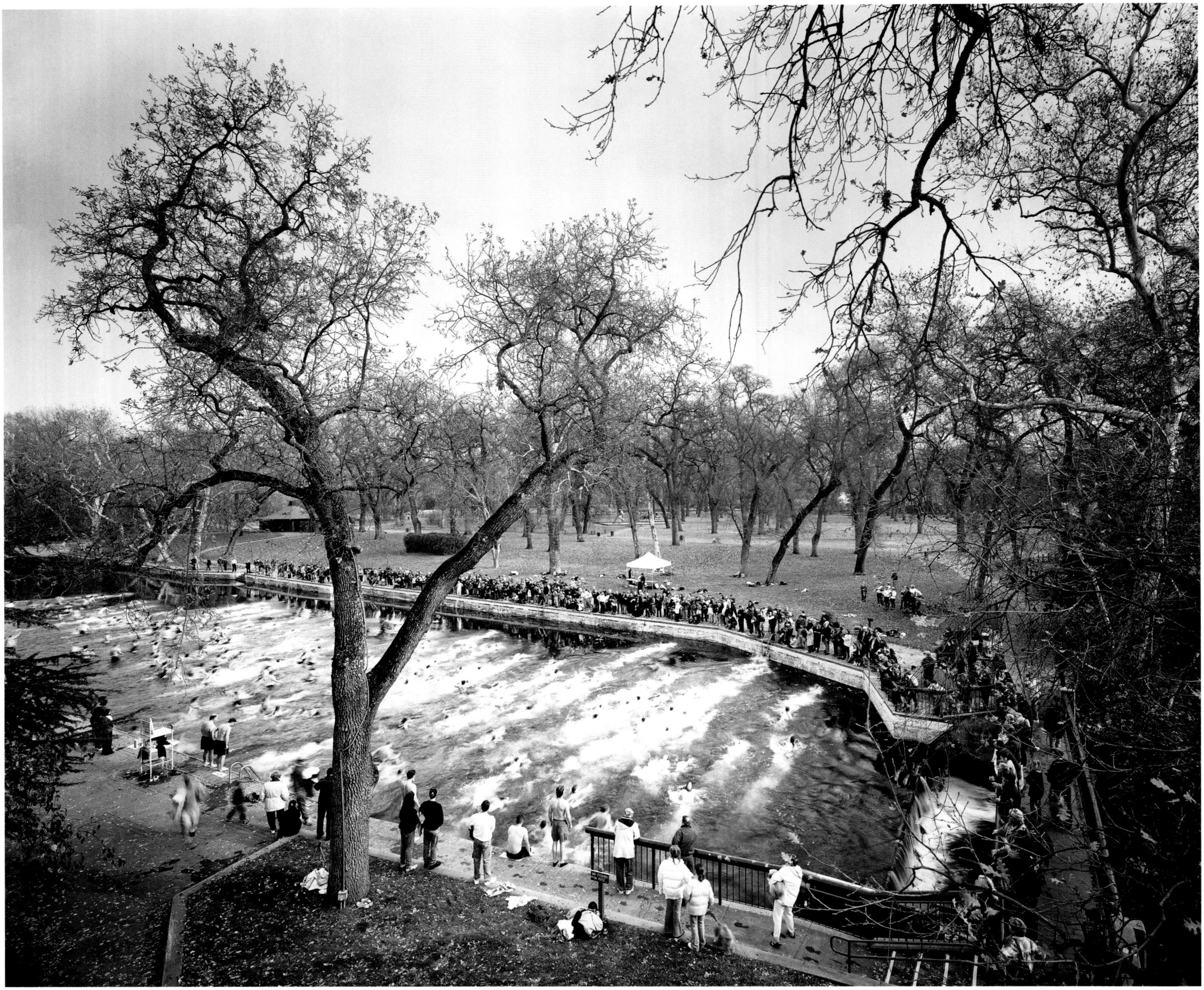

Annual Polar Bear Swim at "One Mile," Chico Creek, 2000

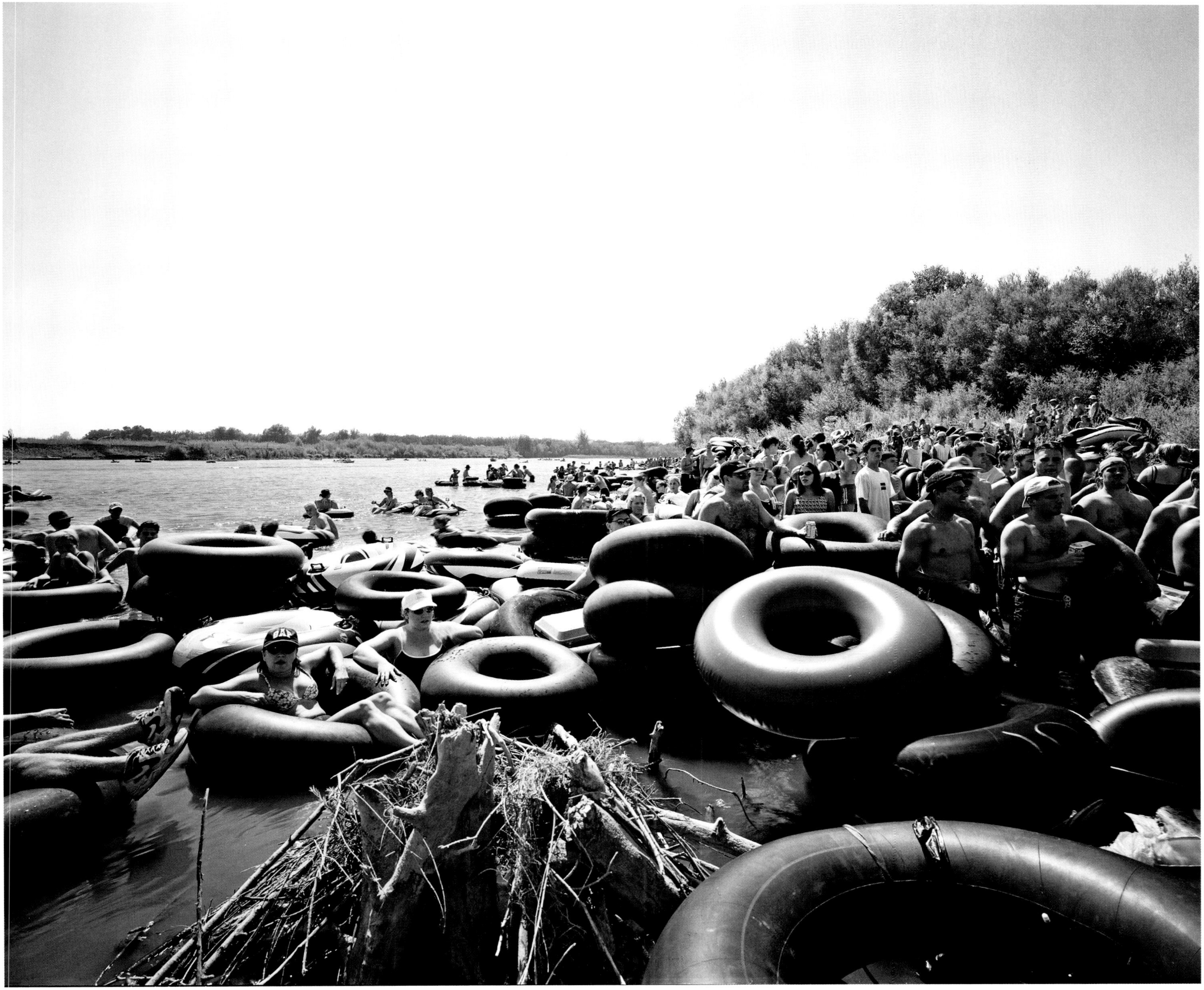

Labor Day on the Sacramento River, 1999

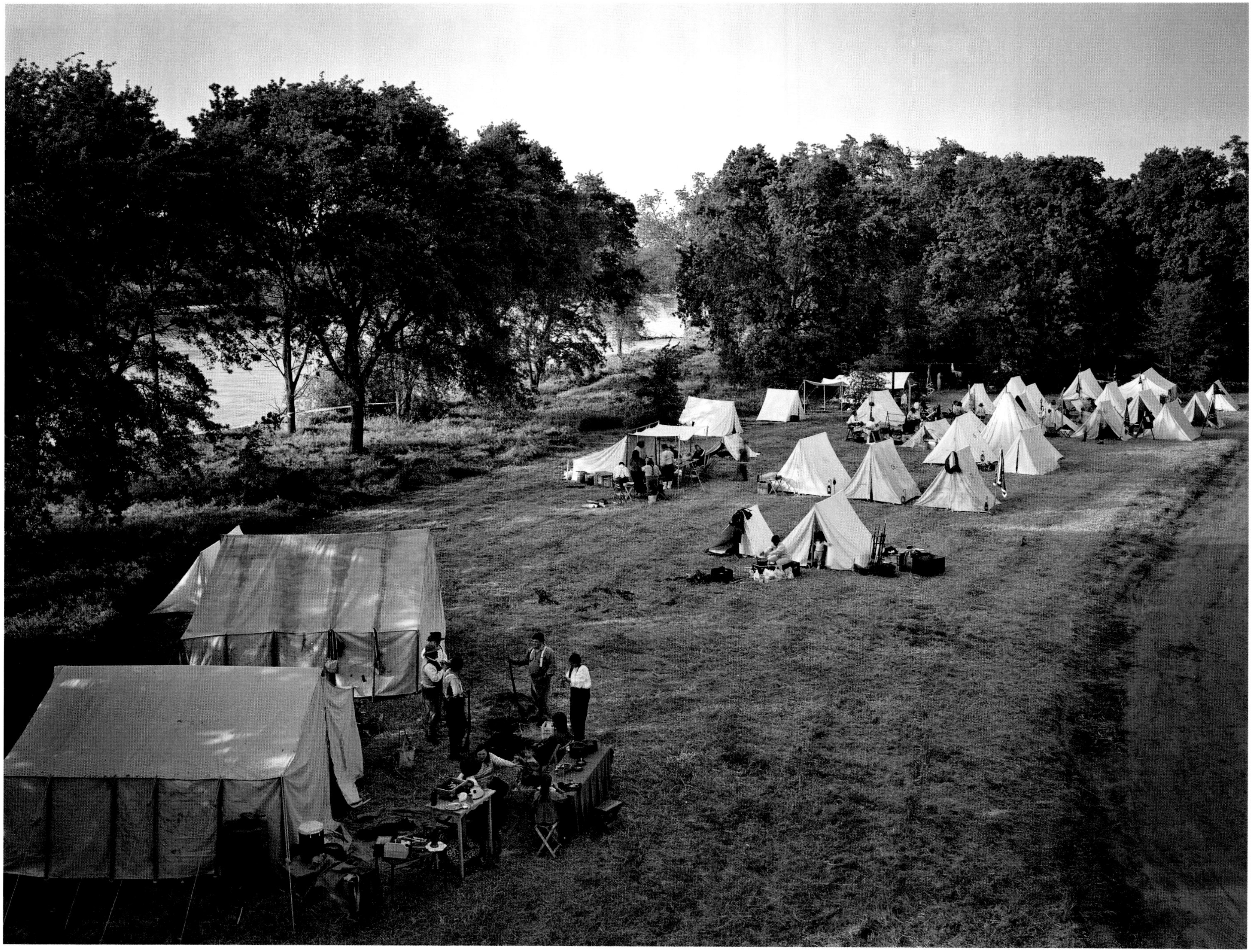

Civil War reenactment at Woodson Bridge State Recreation Area, 1998

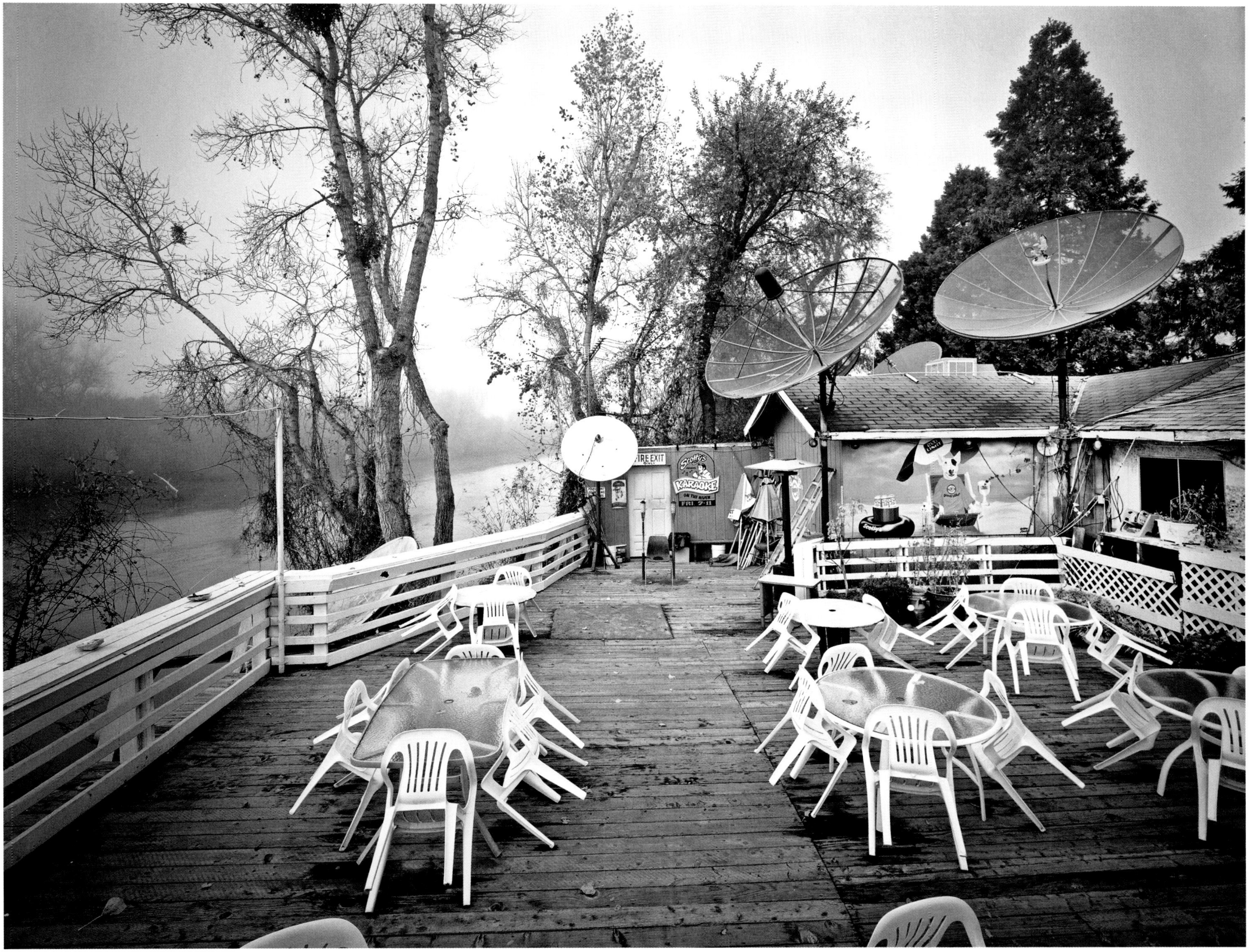

Chairs at Scotty's, 2005

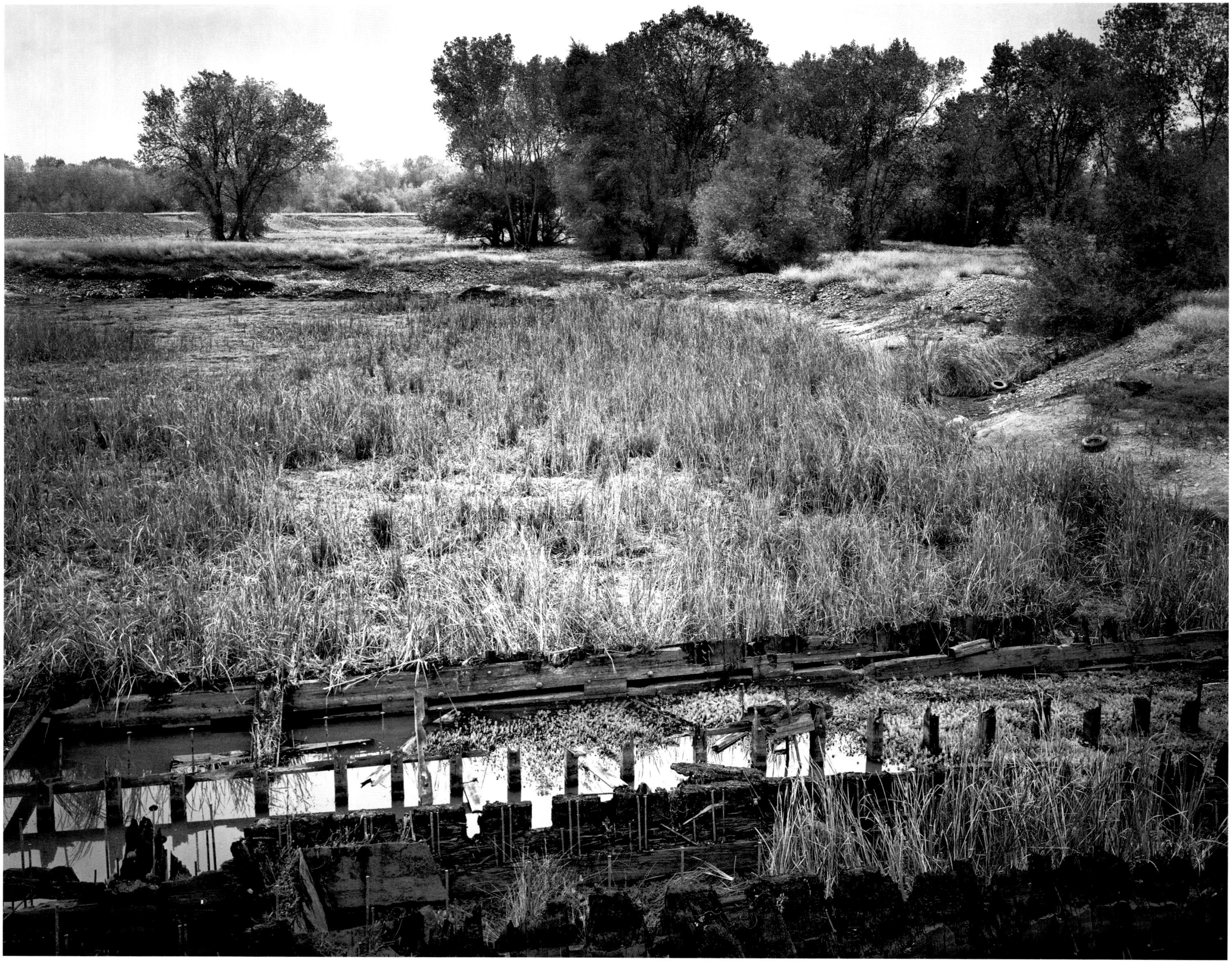

Remains of a gold-dredging operation near Oroville, 1997

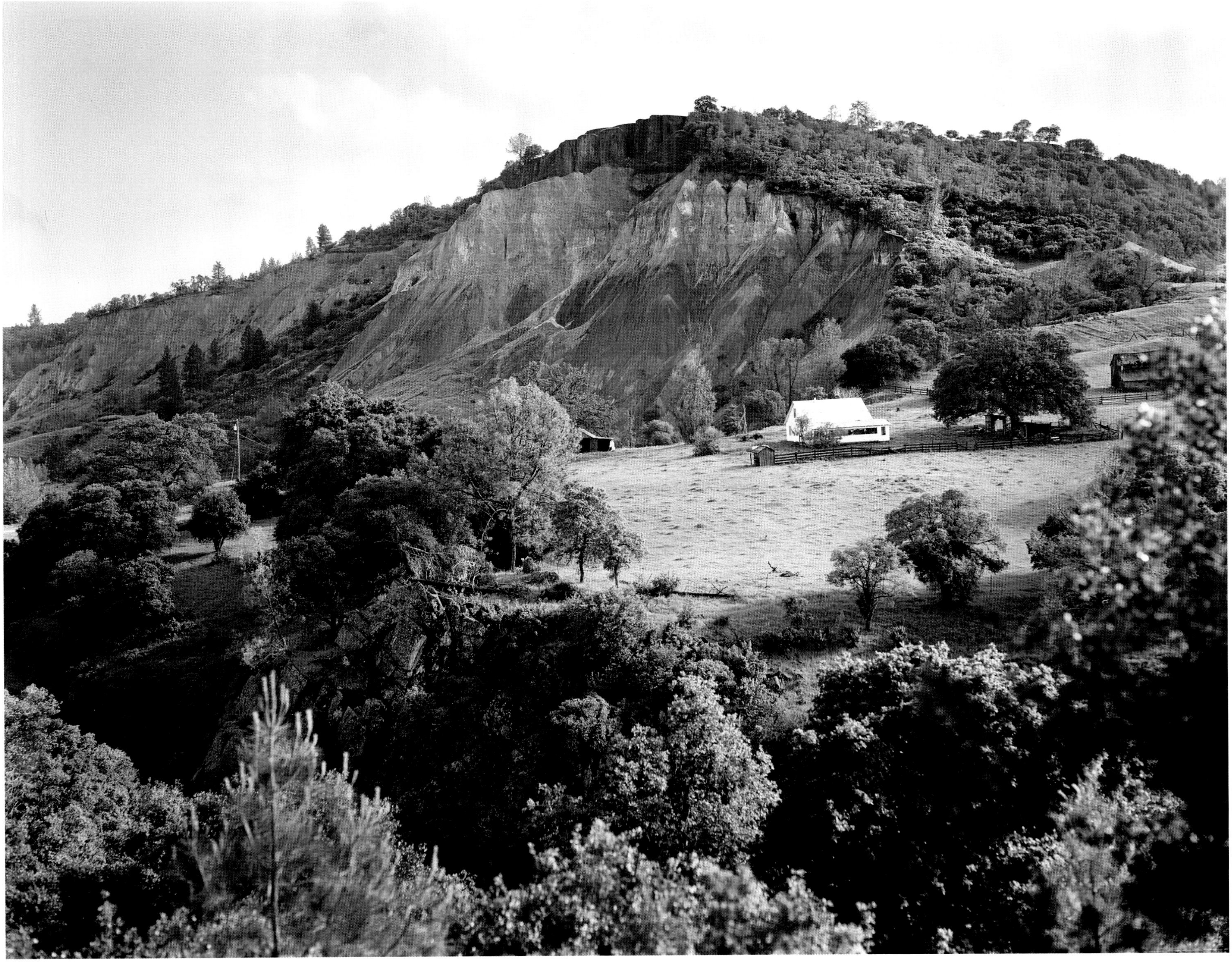

Cherokee Hydraulic Mine, 1998

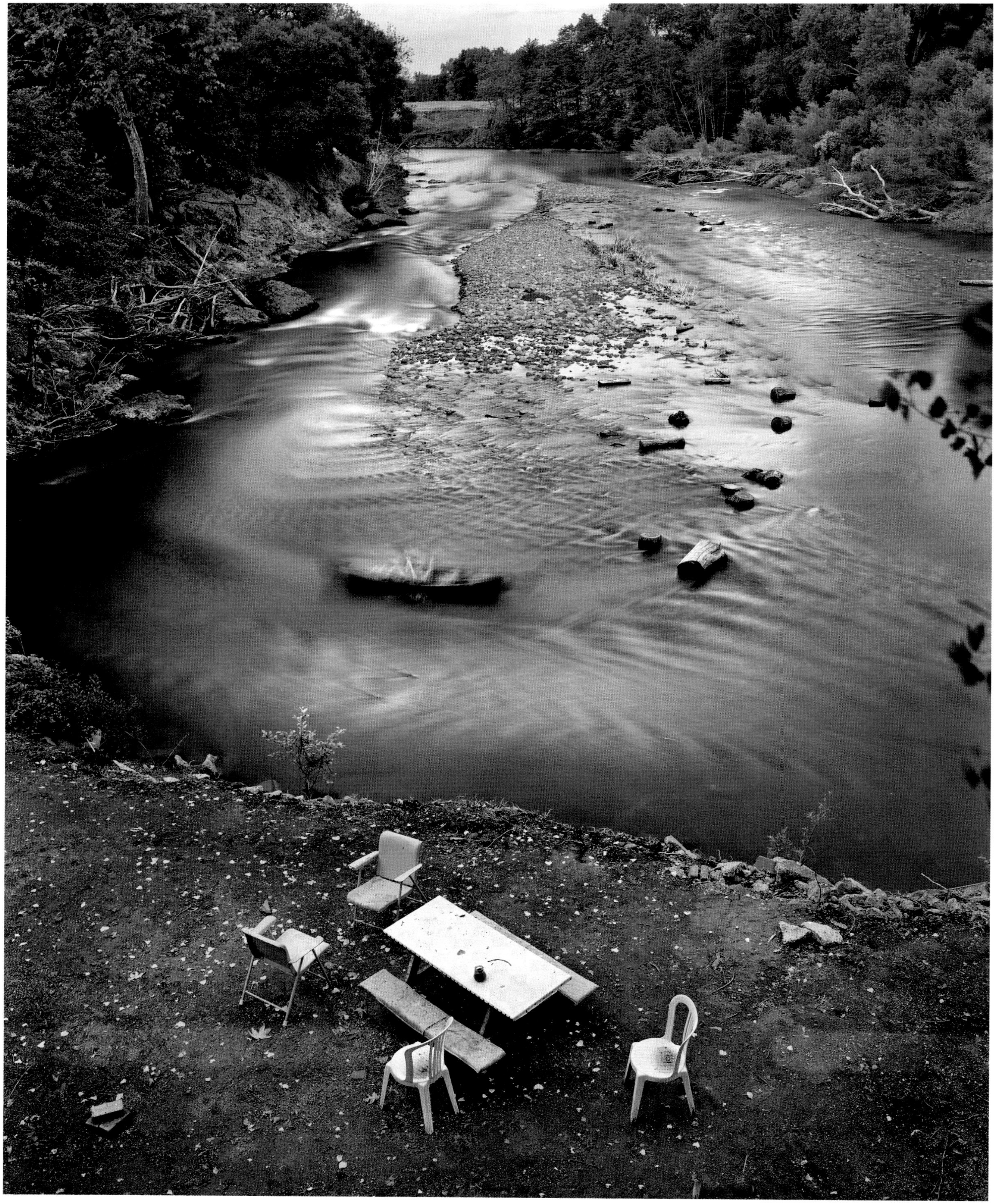

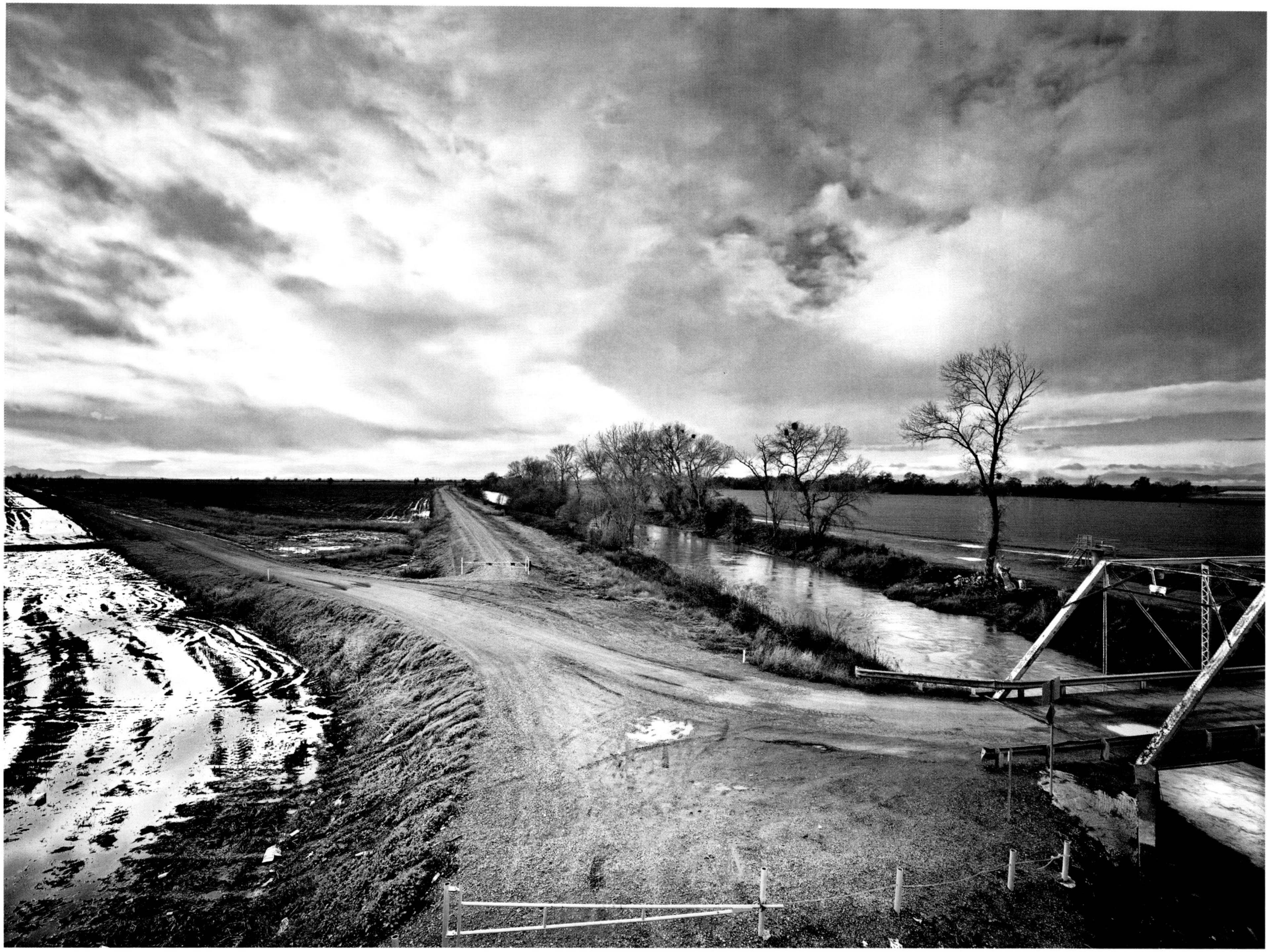

Butte Creek north of Western Canal, Durham, 1996

« Canoe with chopped cottonwood logs (midstream) and naturally fallen trees (sides) on Butte Creek, 1998

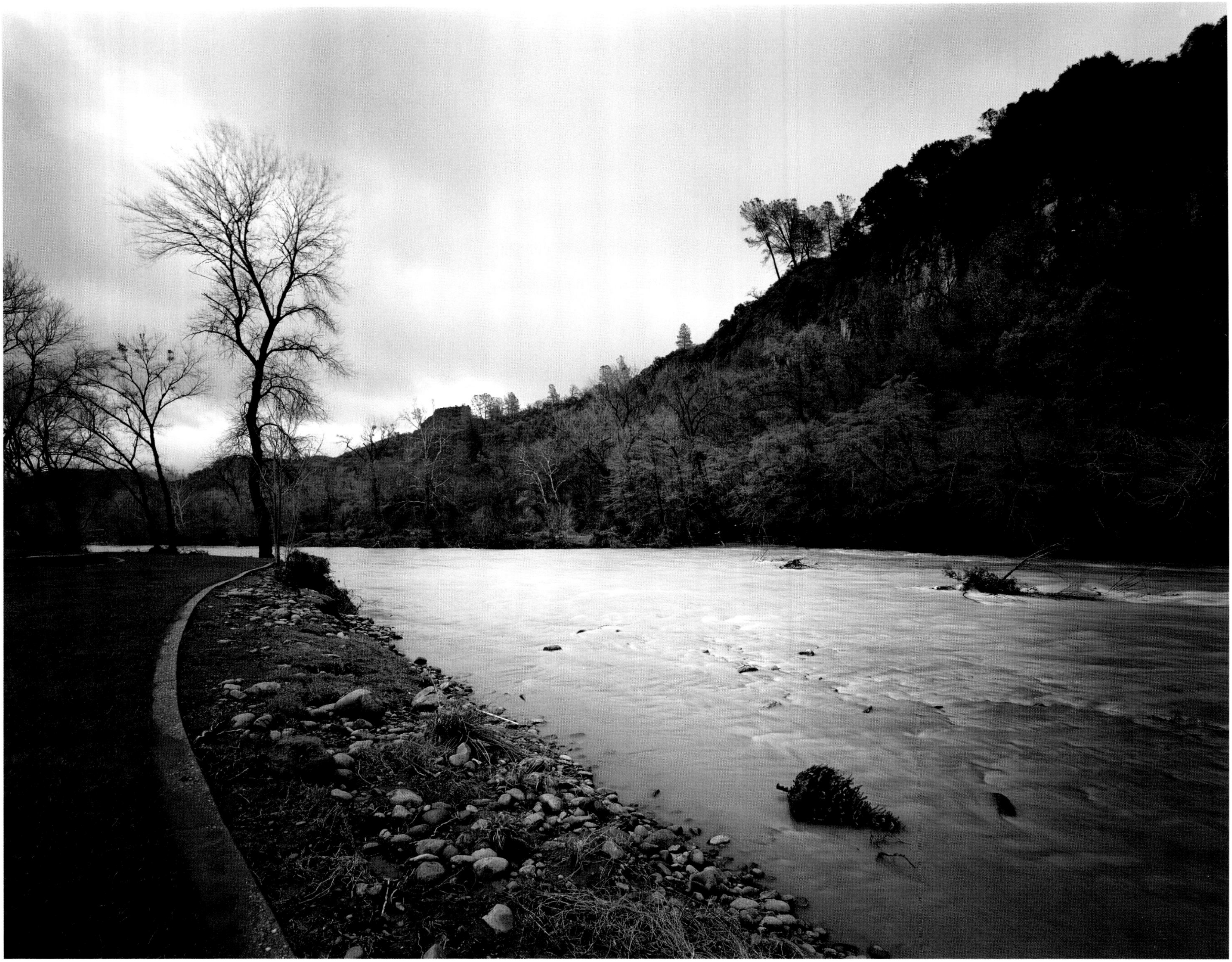

Christmas tree on Butte Creek, 1997

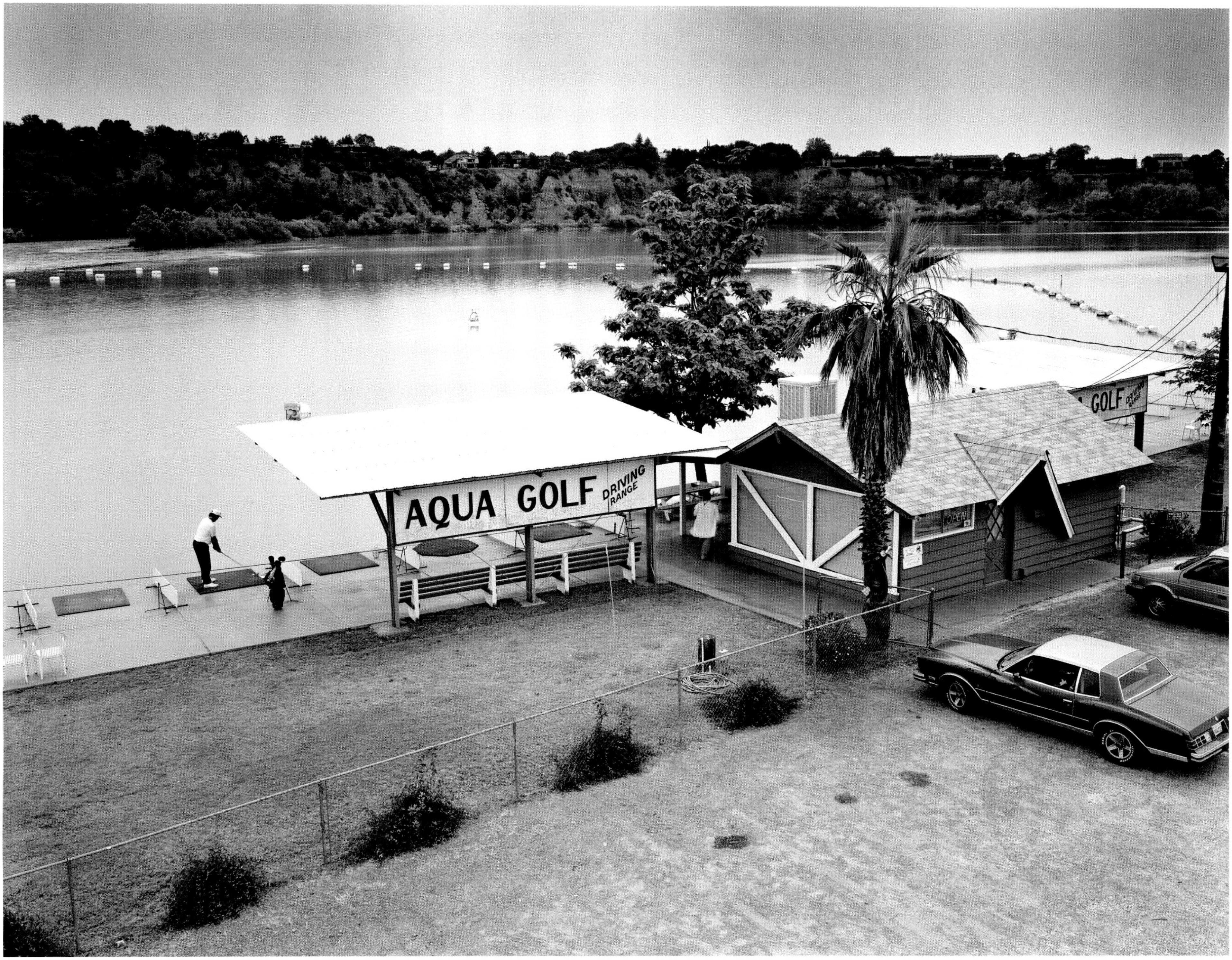

Aqua Golf on the Sacramento River, Redding, 1999

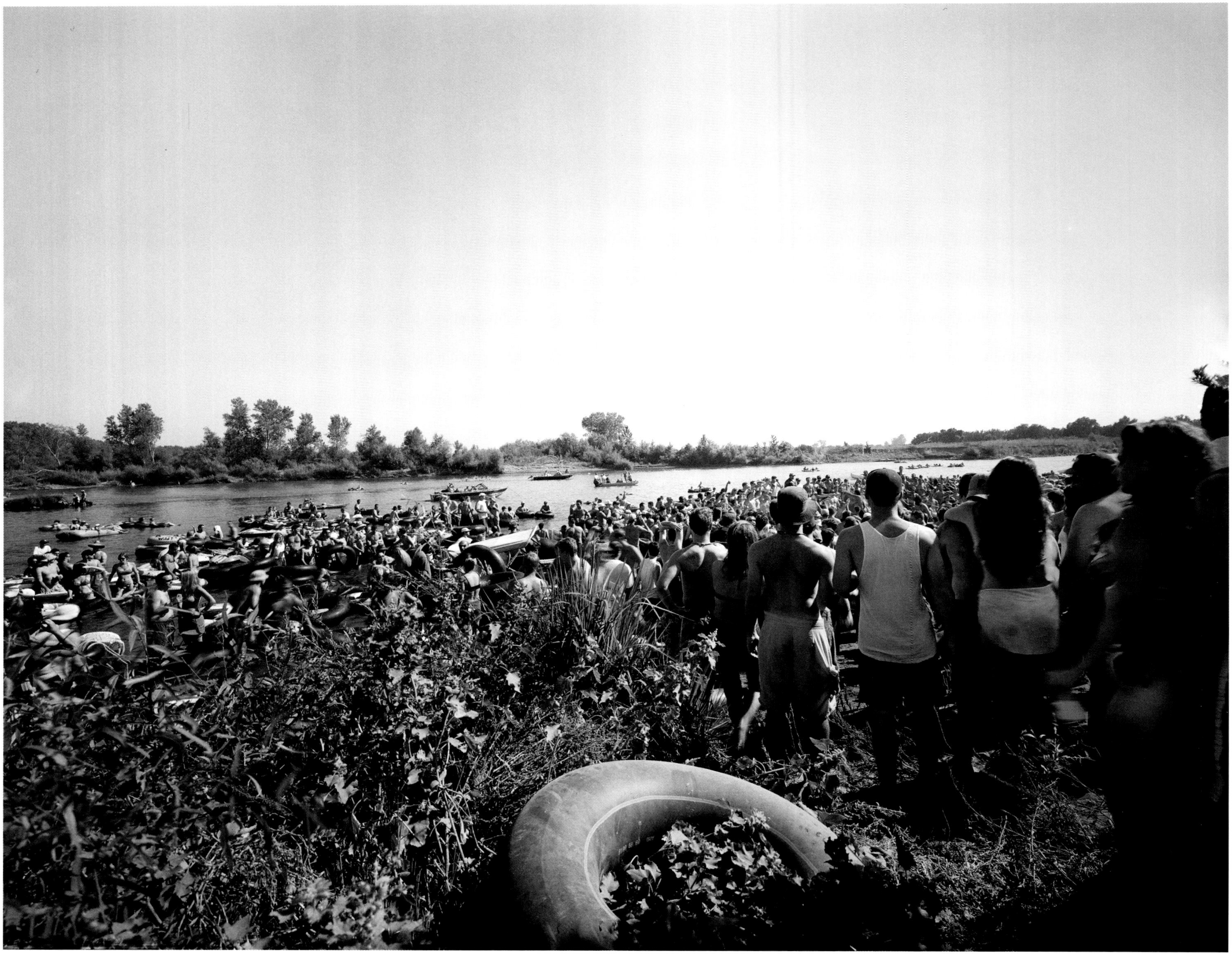

Labor Day weekend near Scotty's, 1999

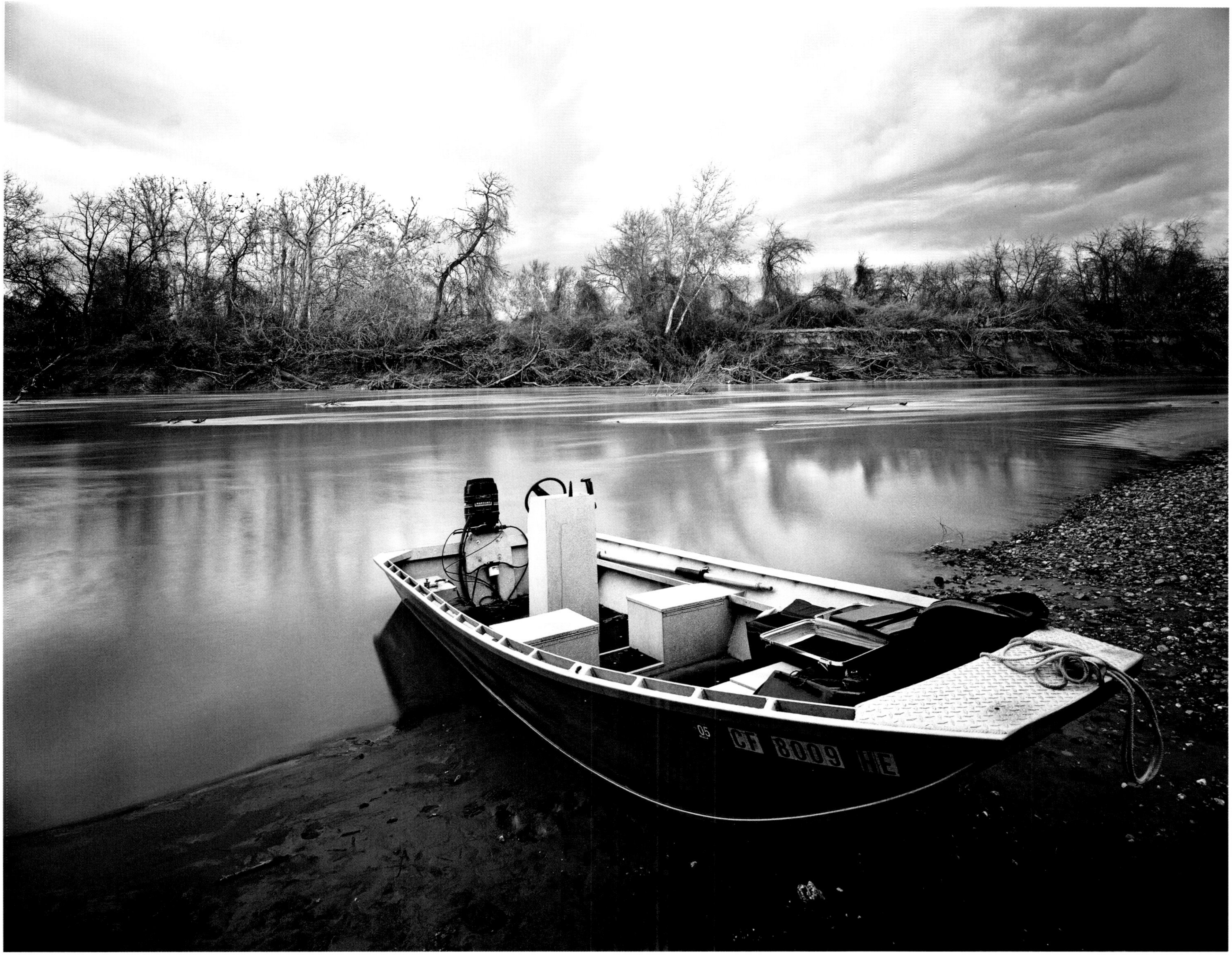

Storm brewing with boat, sycamores, and cormorant rookery just downriver from Ord Bend boat landing, 2003

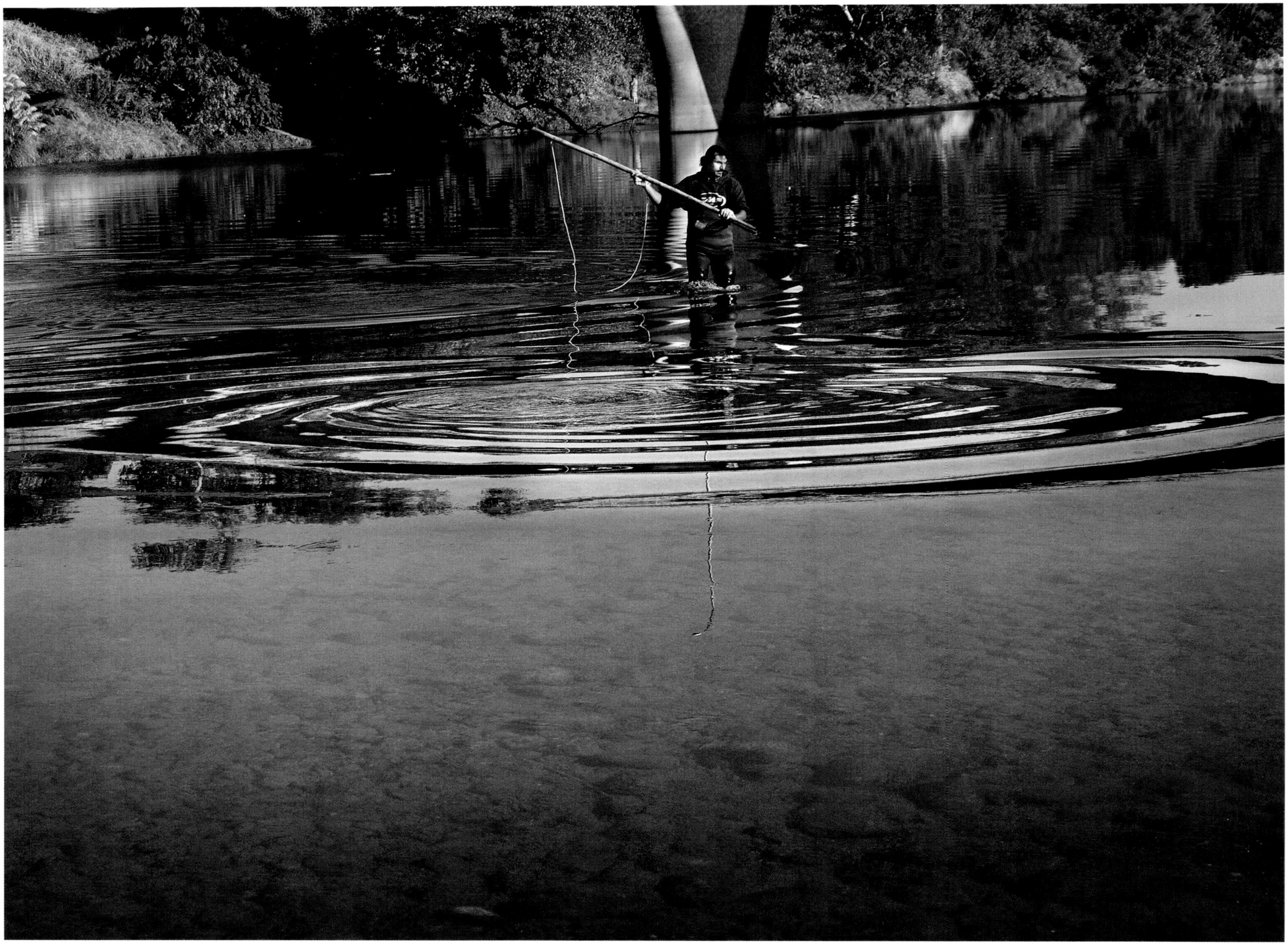

John Clark spearing the first salmon for Oroville's annual Salmon Festival on the Feather River, 2011

Camouflaged duck rig at Chico Landing, 2005 »

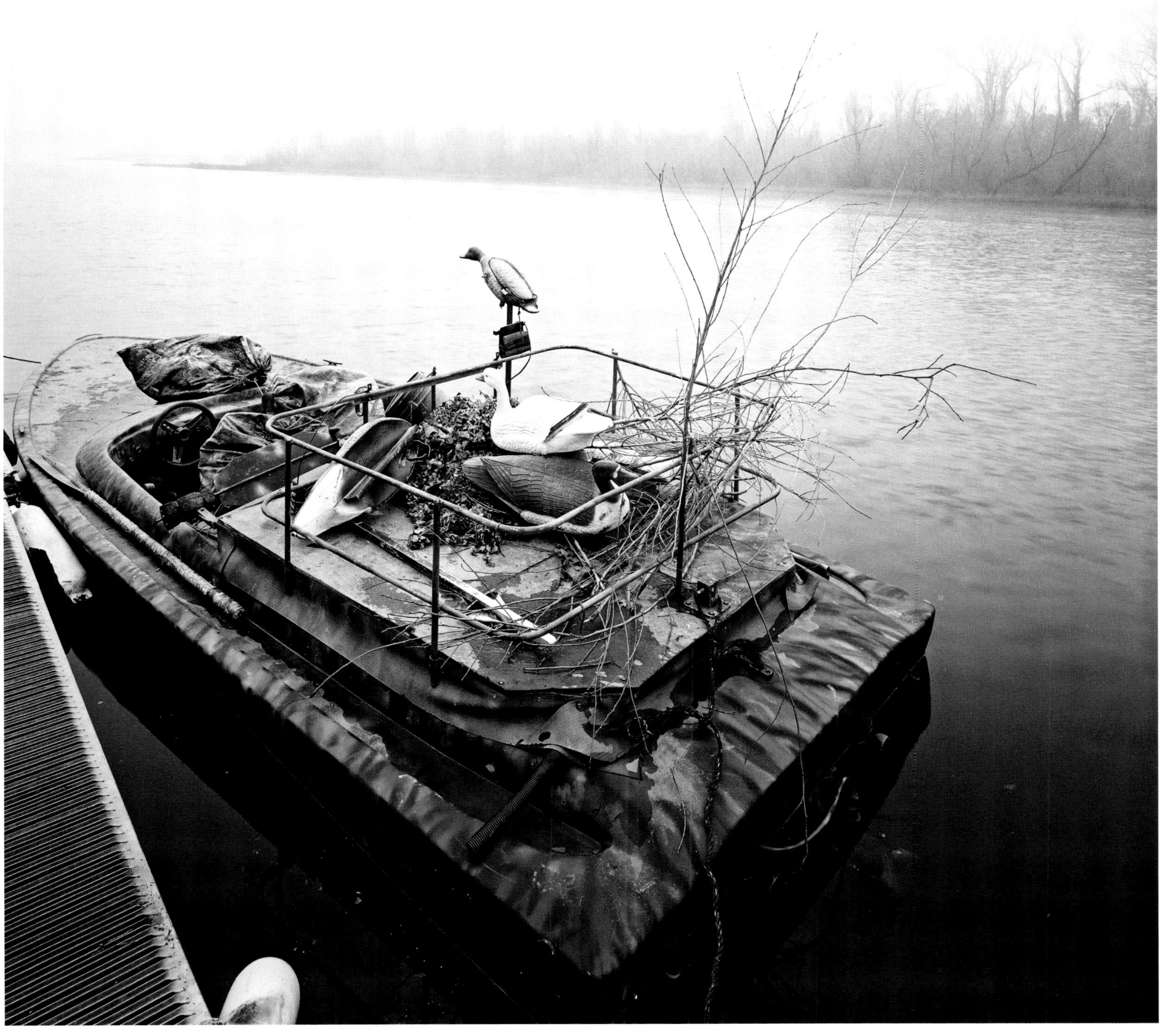

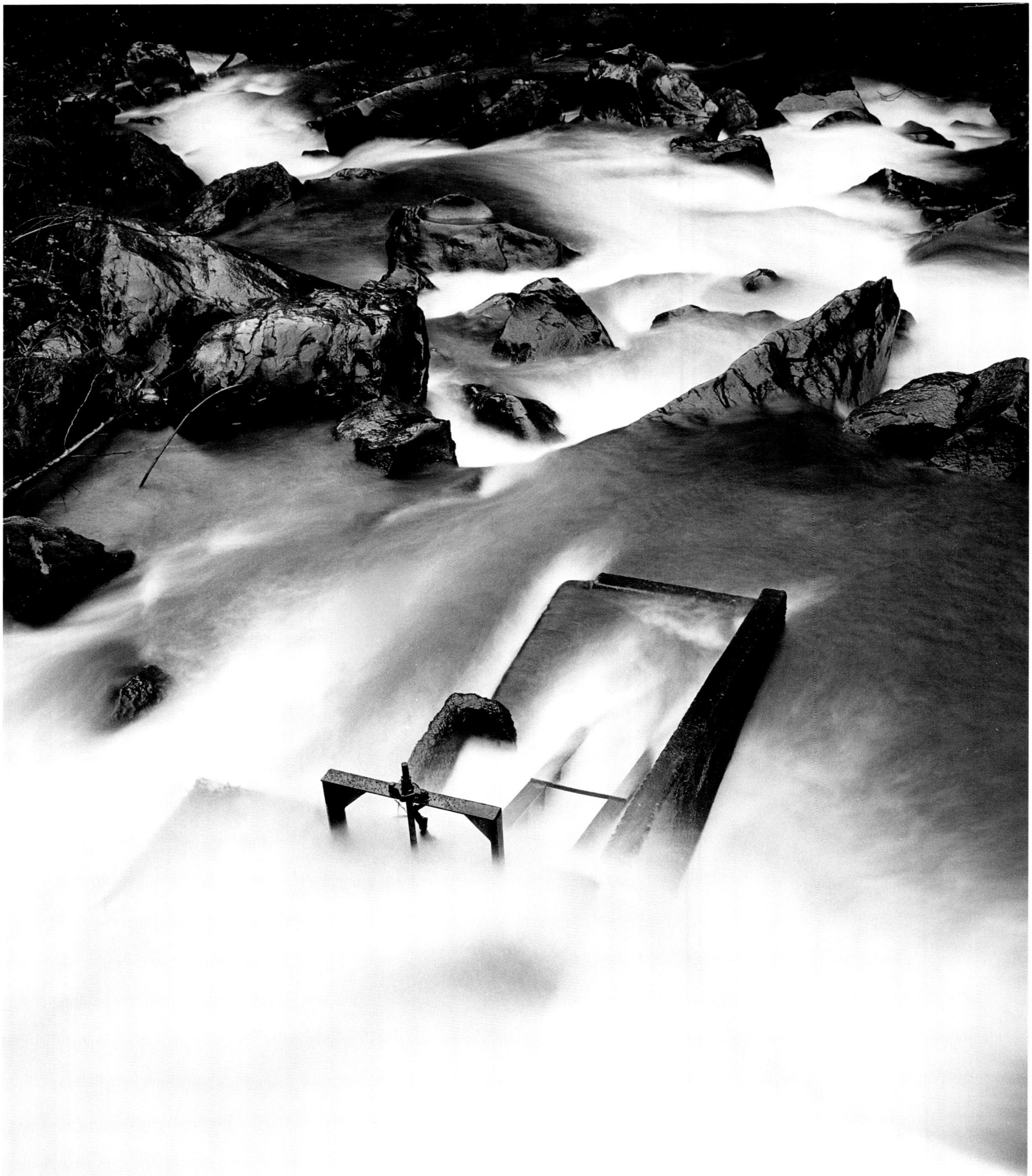

C O N T R O L

SOME SAY WATER is the *new oil*—the substance nations will fight over, die over, and pay each other to dominate. In truth, water is the old oil.

Before we built our lives on petroleum, we centered our villages on water. We knew wherein lay our food, drink, mobility, and spirituality, and we gravitated there. People congregated on the world's major rivers and streams, and their tributaries: Tigris-Euphrates, Nile, Amazon, Mississippi-Missouri, Yukon, Volga, Yangtze, Mekong, and others of great volume and power. Water lay at the center of civilization.

Along with its helical flow and skill for finding weakness, another innate property of water gradually emerged: water is a finite resource. Earth's fresh water in liquid, gas, and solid form totals about 366 trillion gallons or 1,098 million acre-feet (one *acre-foot* equals an acre, an area approximately the size of a football field, one foot deep in water). Although it translates among states, it is neither created nor destroyed; there is no more of the substance available to us that we know of, and no less. Water consistently runs through the hydrologic cycle—that global, sun-driven pump that circles a drop of precipitation from sky to land to sea to sky again—but it is not created anew.

When there is enough of a finite resource to go around, peace prevails. A belly at ease doesn't grumble. Indigenous people of the Sac watershed were less given to conflict than many first nations in North America. The naturally abundant world of the early Sac and San Joaquin has been credited for the scarcity of intertribal warfare in California. Yet the same amount of water that easily sustained the first people could not fully serve the burgeoning population of immigrants from water-rich geographies. Struggle for control of rivers for water-intensive businesses and settlements followed the advent and domination of the Anglo-Europeans. That it was an arid land didn't alter their thirsty ways. Low-impact practices of the past were pushed out; the influence of those who had long been living with the ephemerality of water in the dry climate faded.

River, rival—both words have their roots in the Latin *rivalis,* meaning "to use the same stream as another." Those who use the river for a given purpose necessarily conflict with other users, other lives.

As indigenous people were defeated and driven from the Sac, immigrants who settled the floodplain by the thousands, beginning in the 1850s, soon faced a new adversary: repeated and extensive flooding. Communities protected their farms, homes, and towns by constructing levees, earthen floodwalls paralleling the river. Because of the nature of water, building the bank higher on one side of the stream simply spiraled flow toward the other, where banks were often less fortified. Landowners farther downstream, too, lost land when

« Slide-gated water-control structure on Battle Creek, 1995

levee-constricted flow attacked and undermined other, unprotected embankments with increased intensity. Raising riverbanks simply required that more banks be built up in response.

Levee construction continued with a commitment well known to landowners threatened with loss of the earth under their feet. First efforts began locally, with private parties building initial berms around three feet high, then higher and wider. With time, great concrete walls were erected, in some places to twenty-five feet high and two hundred feet wide at the base.

The Sac's levees today follow the river in some reaches like twin snakes hugging the shore. In many places the floods that seasonally replenished soil fertility and ecosystem health are now isolated from the valley floor.

Levee building brought another adversary as well: channel flushing. Hydraulic mining for gold had washed tailings into the Sac's tributaries in enormous masses, filling the channels first with fine sediment, then coarse. With levees constricting floodwaters from going overbank, higher and quicker instream flows carried debris faster and farther downstream. Pulses of gravel-heavy loads migrated from channels to farmers' fields, rendering the soil useless for agriculture. River channels that had been navigable by steamboat choked, first in the upper reaches, then the lower. The clogged riverways lost more and more natural flood capacity, overtopping with increasing frequency and ease. Once a flooding solution, levees began to send complications throughout the watershed, as the touch of a hand ripples a pool.

Fight for control of the Sac moved indoors. In 1884, an early environmental ruling came down from the federal court to California, holding that discharge of debris such as mine tailings creates irreversible land and water damage. The ruling halted hydraulic mining on the Sac and tributaries, ending that particular chapter of contention among rival stream users. Still, it would take decades for gravels to flush through to the bay-estuary. Some riverbeds will not be clean of the debris for generations, if not geologic lifetimes.

Entering the 1900s, court struggles focused most frequently on *rights:* who owns water and therefore determines its fate. The English common-law tradition of riparian rights—ownership by whoever owns the adjacent lands—had immigrated to California with many Europeans. Riparian rights conflicted, however, with appropriative rights—ownership of water resources by whoever used the water first, regardless of location relative to the stream. Lawsuits became the order of the day as Californians grappled over the Sac and other rivers: in 1922, the California Supreme Court was hearing more cases on water decisions than any other issue.

Over time, riparian rights were judged to take precedence; to this day, they usually prevail over appropriative rights. But California's water use has always been and remains a complex composite of the two traditions, and neither on its own holds the answer to best use of a common resource. Water use in a state in which the bulk of water resources are consolidated in the north grew more complicated as population spread throughout all regions, regardless of supply. As early as the 1870s, plans were drawn to transfer water from the Sac to semiarid tracts of land in the south Central Valley. Ambitious plans were proposed, then deemed to be too expensive, then partially implemented, then abandoned. Finally, two proposals for statewide projects emerged: beginning in 1933, the Central Valley Project (the federal Bureau of Reclamation's updated version of an earlier flood-control project) and, beginning in the 1950s, the California Department of Water Resources State Water Project.

The answer to who would prevail in controlling the river was *both.* State and federal governments would cooperate in a heavily negotiated and debated collaboration to see the projects through. Today the Sac's flood-control and power-generation structures are embedded in the fabric of the landscape. Elements of the Central Valley Project and the State Water Project relate in almost unseen ways.

Broadly speaking, the federally funded components of the Central Valley Project north of the Sac–San Joaquin confluence are found on the Sac's main stem and eight tributaries, including the Trinity and American. The project includes Shasta Dam, fifteen other dams, and nine aqueducts and canals. The largest beneficiary of the Central Valley Project is the Westlands Water District serving the San Joaquin Valley, far south of the Sac watershed.

The California-funded components of the California Department of Water Resources State Water Project center on the Feather River and San Francisco Bay watersheds, comprising five major dams, including Oroville and San Luis (a joint federal-state facility), and the California aqueduct. The Kern County Water Agency in the San Joaquin Valley and the Metropolitan Water District of Southern California purchase the greatest portion of State Water Project water.

Both projects serve users all over the state, but because 75 percent of water resources are in the northern half of California and 80 percent of water demand occurs in the south, the scale of weighing water purchases against receipts tips away from watersheds of origin. Overall, the state's combined water systems handle thirteen trillion gallons (39 million acre-feet) of water annually. Such massive agricultural and urban withdrawals result in less than half of the fresh water carried by either the Sac or the San Joaquin River reaching the bay today.

One underrecognized aspect of California's water projects is their effect on underground water, or groundwater. Surface water—rivers, lakes, marshes, and ponds—surface water—is naturally connected to underground supply: surface water is the expression of often larger reserves of fresh, underground water, revealed aboveground the way icebergs show only their tips. Groundwater moves slowly through the matrix of the subsurface geology; in many types of rock, water is only minutely mobile. Precipitation that fell long ago and permeated inch by inch may be stored deep; younger supply makes up shallower reserves. Pumping groundwater draws it through the aquifers, underground geologic layers that store it. Draw on shallow groundwater, and you likely also draw on the supply feeding a river. Draw down deep groundwater, and you eliminate your reserves.

Many landowners in Northern California possess one-hundred-year-old surface-water rights along the Sac River. These landowners are allowed to resell surface water; many take advantage of the resources brought to them by publicly funded systems and resell water to users outside their watersheds—notably south to the San Joaquin. To replace the surface water that landowners have basically reallocated, they irrigate with increasing amounts of groundwater from beneath their parcels. Property owners are also groundwater owners:

they have rights to the water flowing under their feet. In essence, the largely invisible practice of reselling surface water inextricably links groundwater—our final source of fresh water—to the aboveground aqueducts and pipelines of surface-water projects. Groundwater is therefore rerouted through this loophole, and the profits go to the practitioner. The result is the publicly assisted, unaccounted-for, and largely unknown use of groundwater for irrigation, decreasing hydrostatic pressure in underground supplies that water critical ecosystems, reducing flow in streams that are being managed on the basis of surface water allocations, and threatening rampant land subsidence that occurs when aquifers are overtapped. All are offenses against the public trust, for the enrichment of a few.

As the struggle for control of the Sac continues, artifacts of storage and diversion settle into the landscape like magnificent monuments to human hubris. Someday to become archeological discoveries, these installations—the concrete walls of Shasta Dam, the wooden raceways of raised mining flumes, the hand-placed cobbles of the first levees, and the painted wallboards of fishing resorts—are only slightly less ephemeral than our own lives. The river floodplain, with its insistent power of renewal, inherits them all.

Eagle Canyon Diversion canal on Battle Creek Dam, prior to installation of improved fish ladders and fish screens, 1995 »

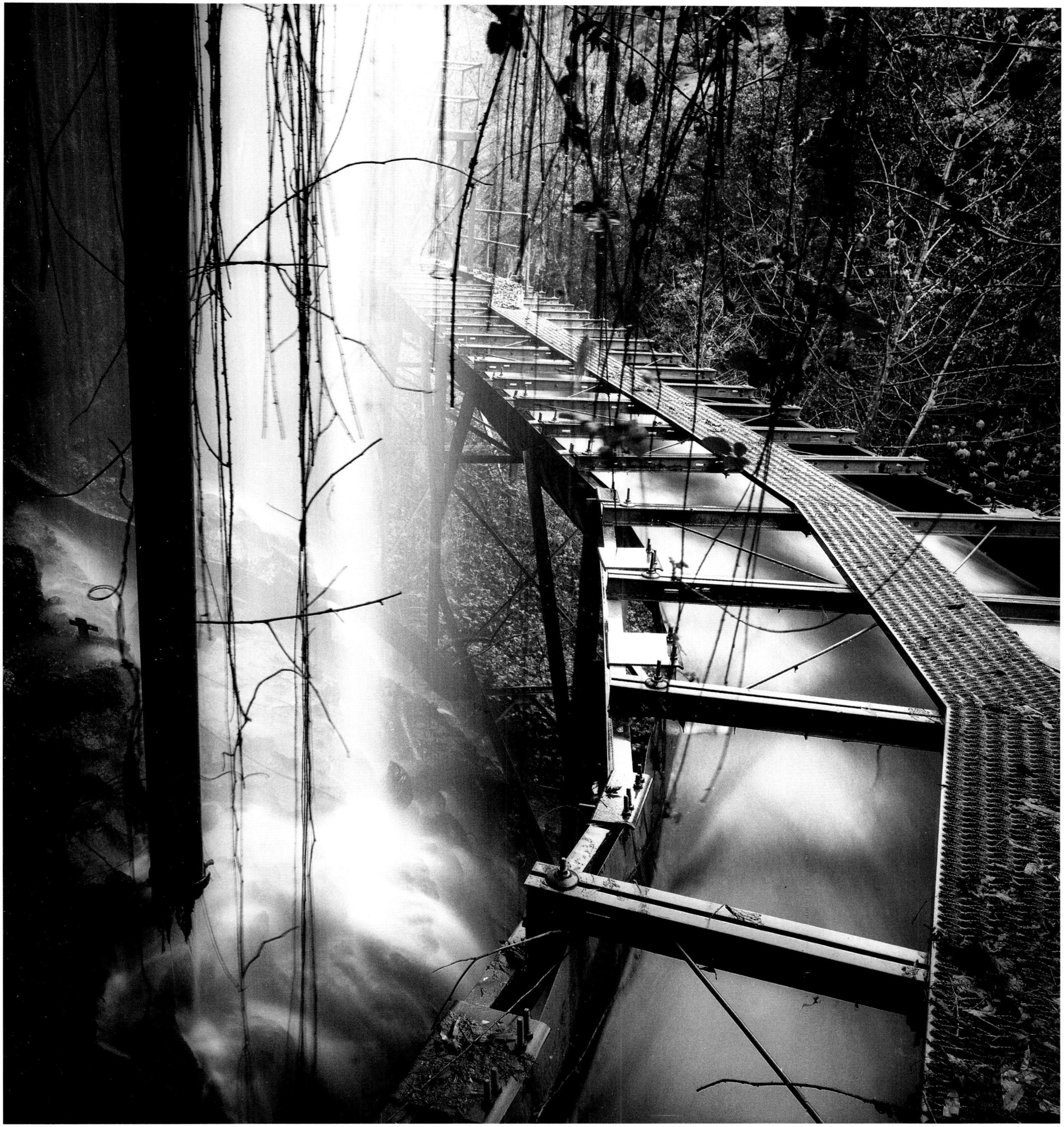

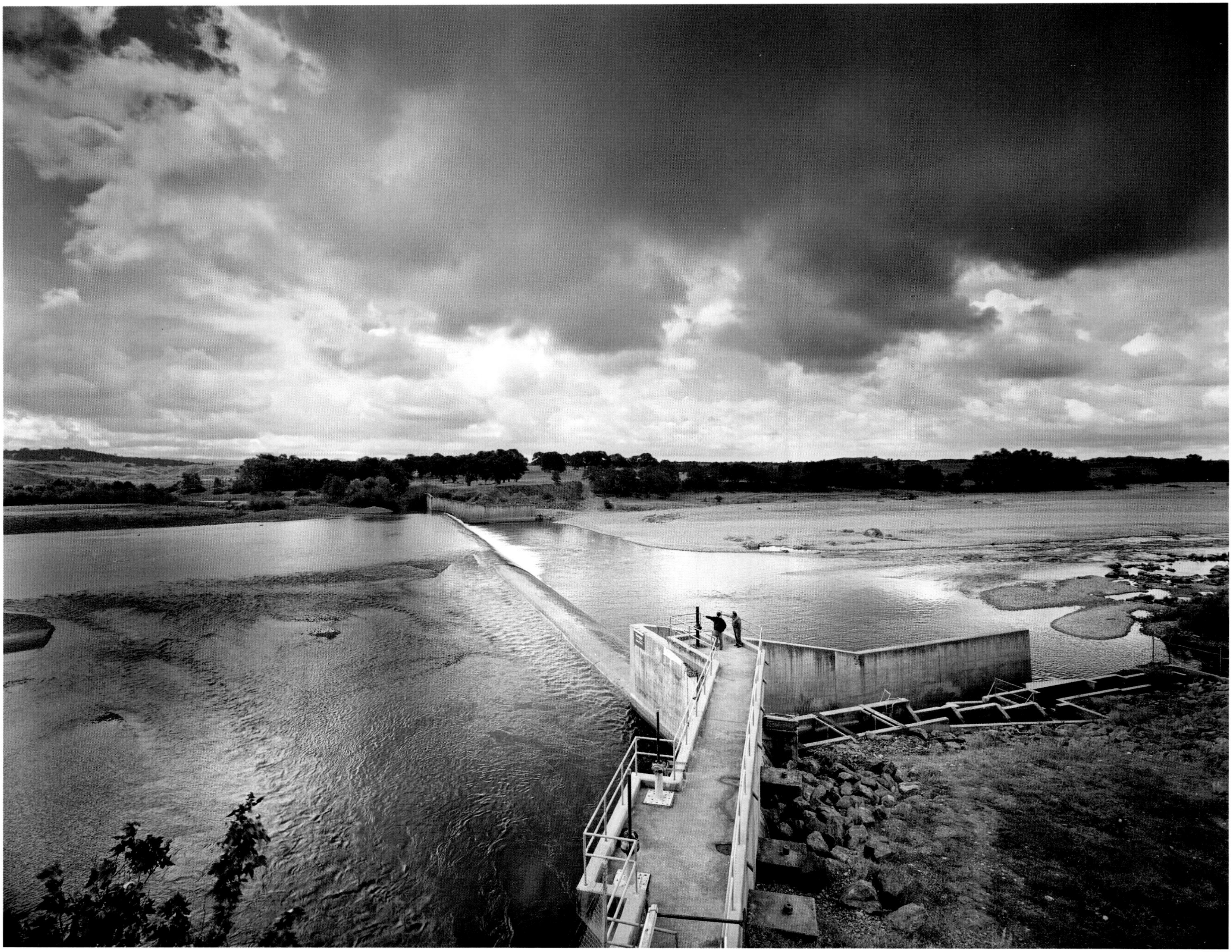

Daguerre Point Dam, designed to stop mining debris from contaminating farmlands downstream, Yuba River, 1998

Cone-Kimball Diversion on Deer Creek, 1996 »

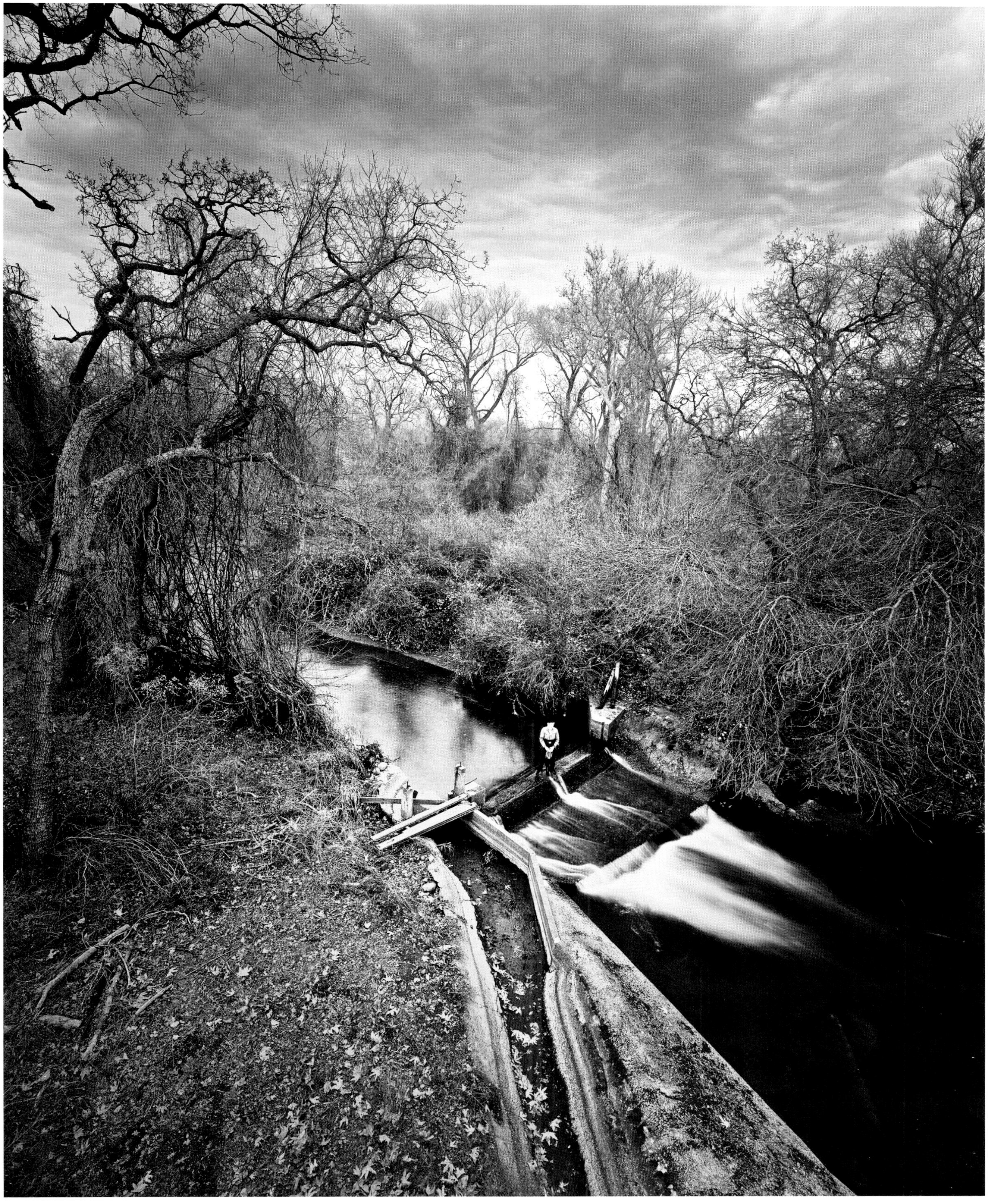

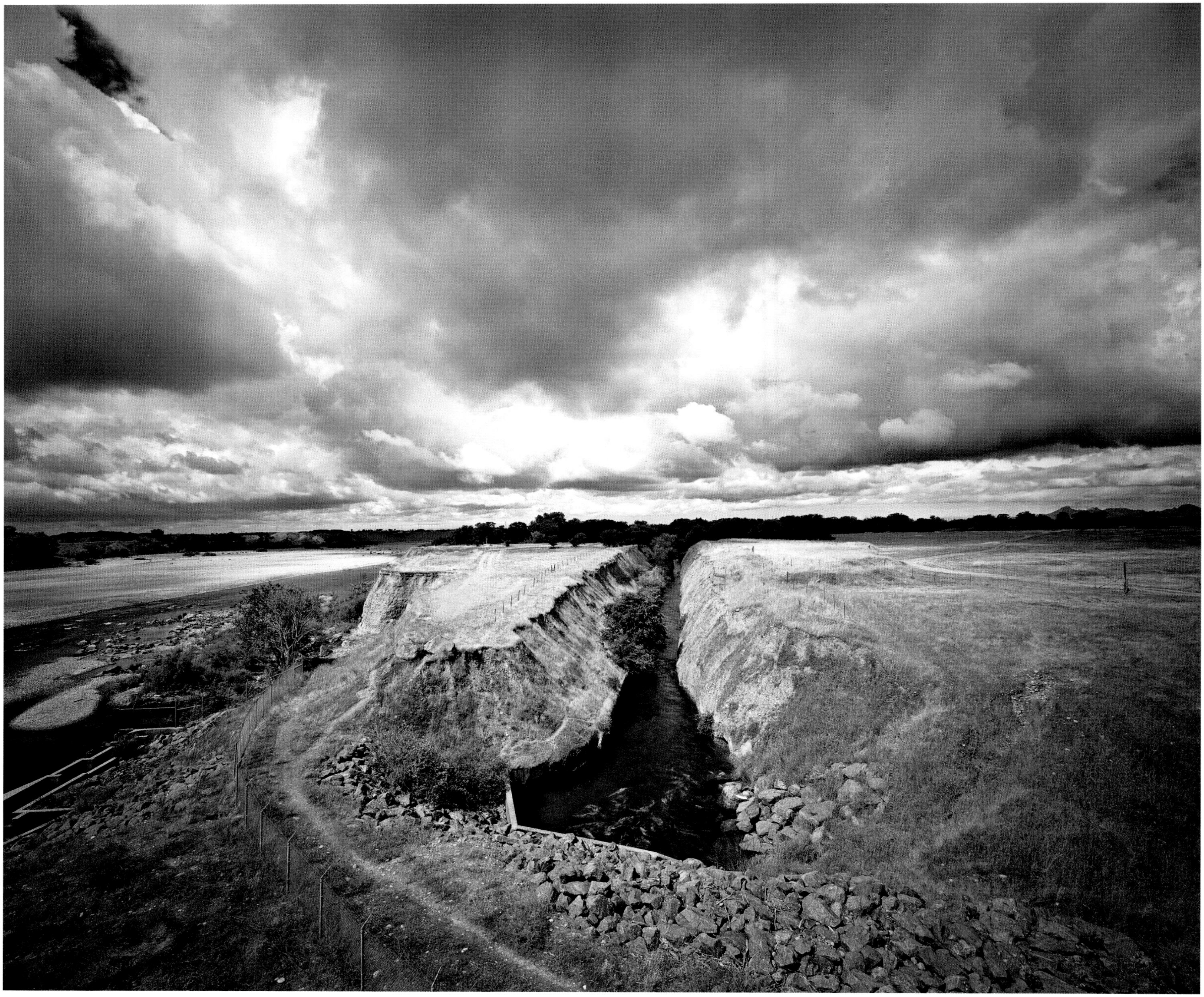

Irrigation ditch at Daguerre Point, Yuba River, 1998

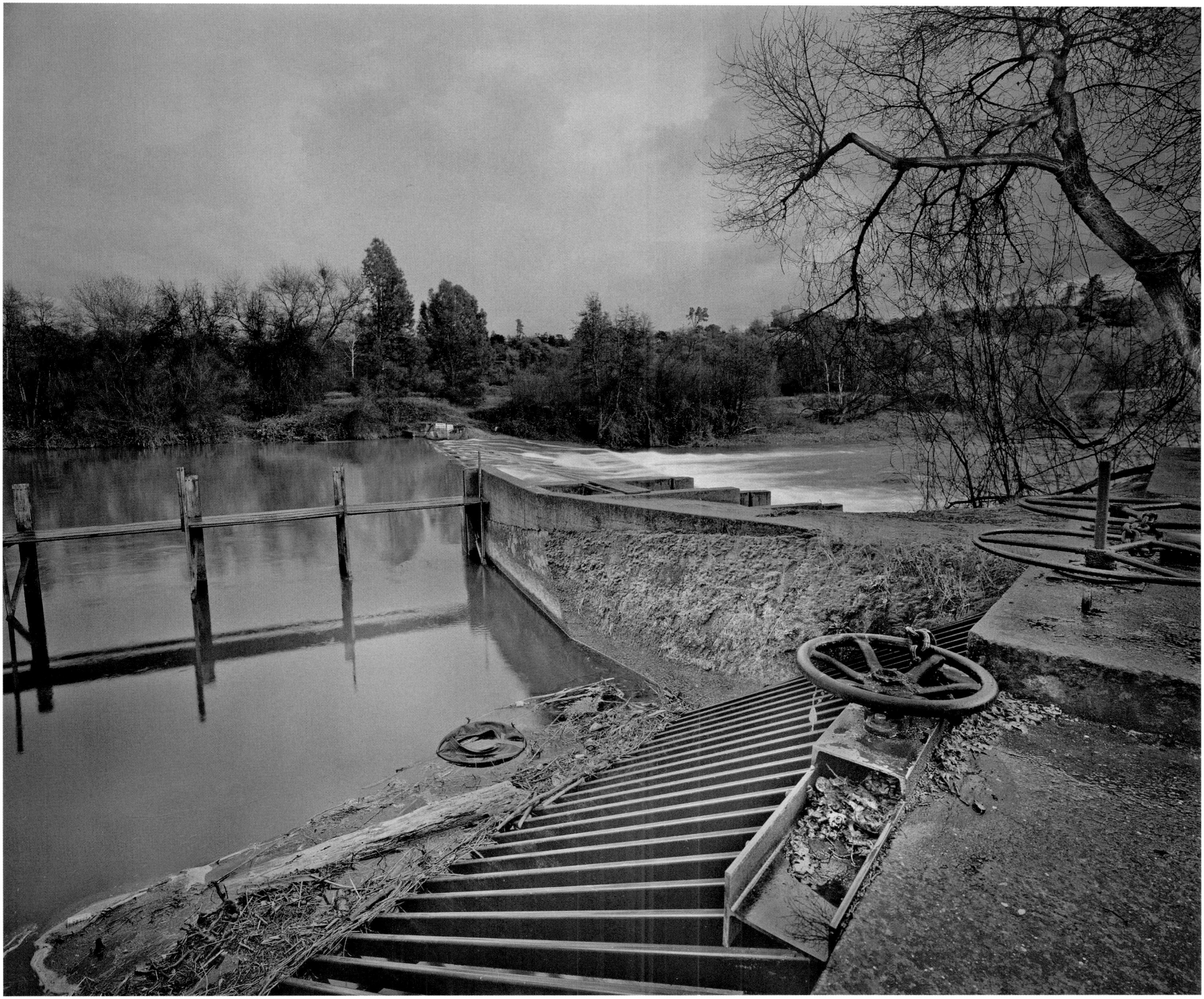

Parrott-Phelan Diversion Dam (also called "Okie Dam") on Butte Creek, 1995

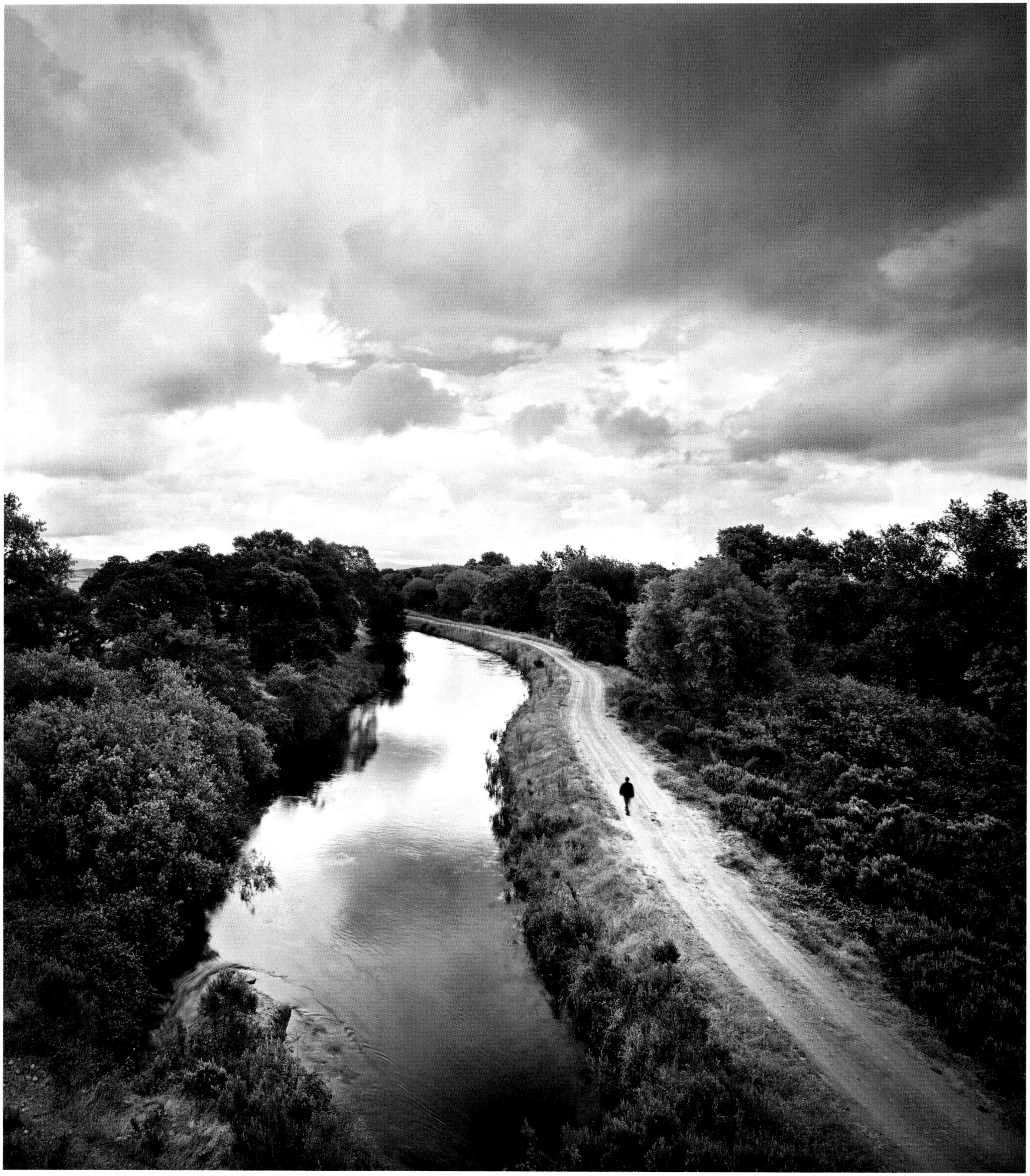

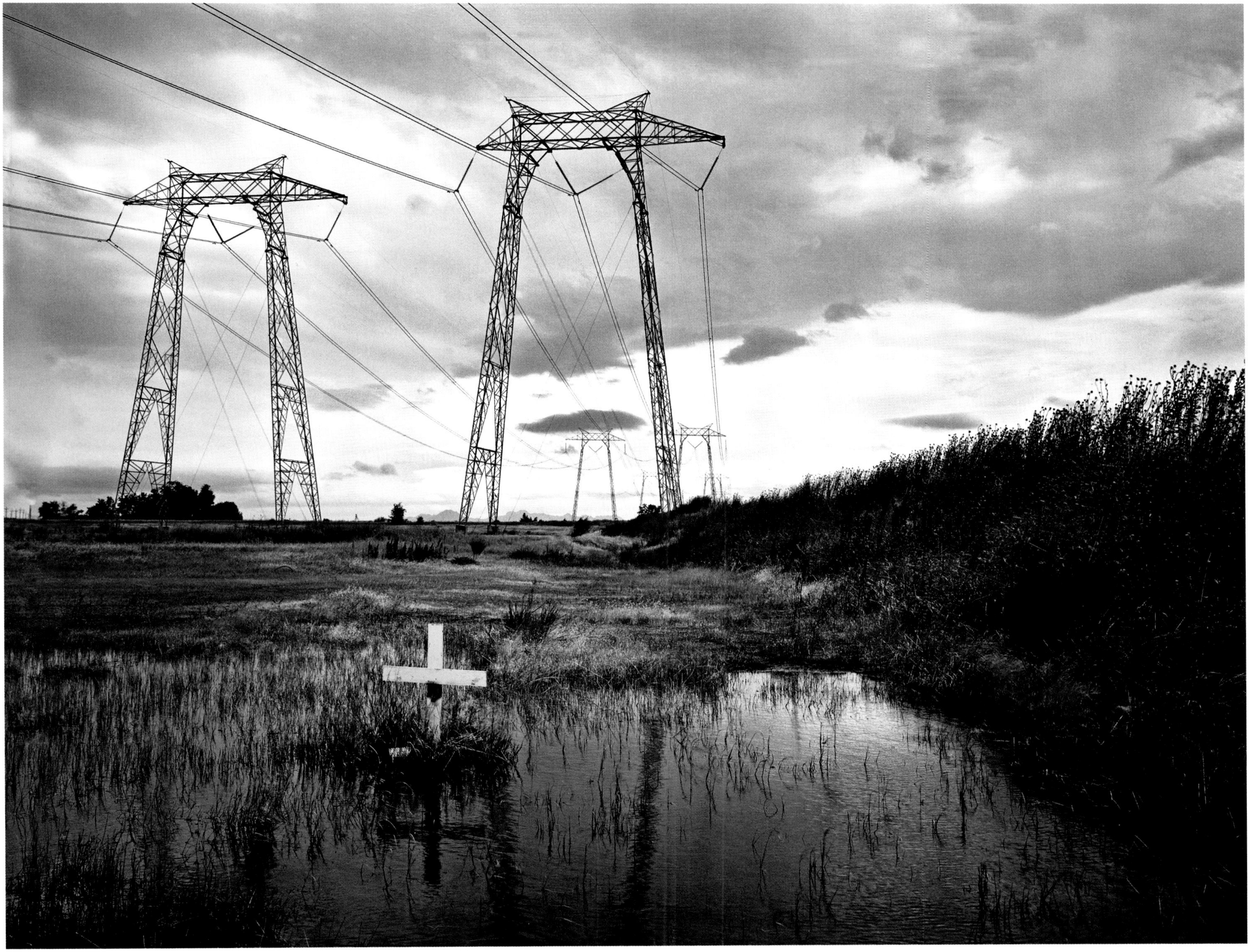

Hydroelectric lines at Highway 99 and Western Canal, 1990

« Daguerre Point Dam diversion ditch for agriculture, 1998

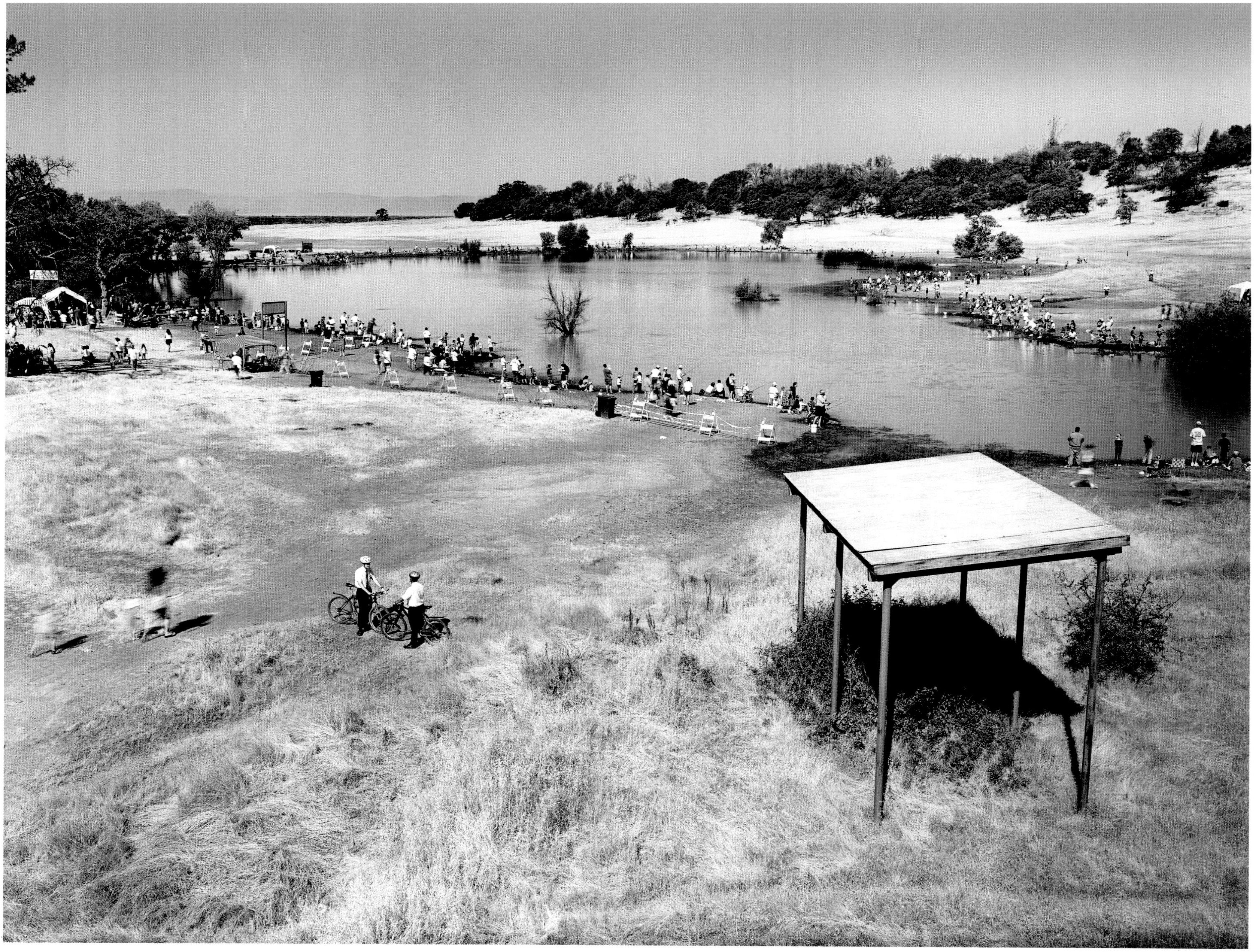

Hooked on Fishing (Not on Drugs) derby at Horseshoe Lake, 2003

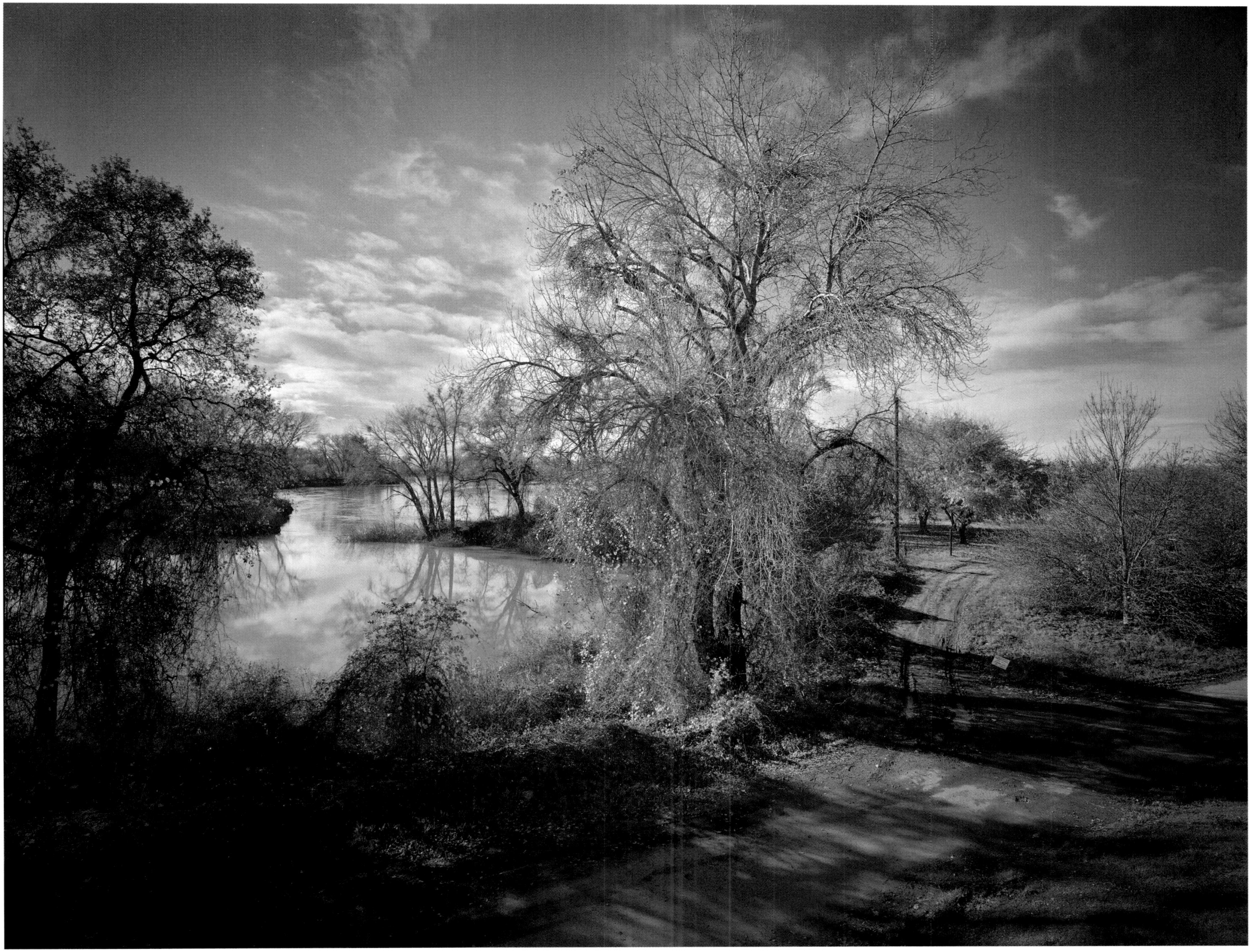

River-ranch interface near Colusa, 2000

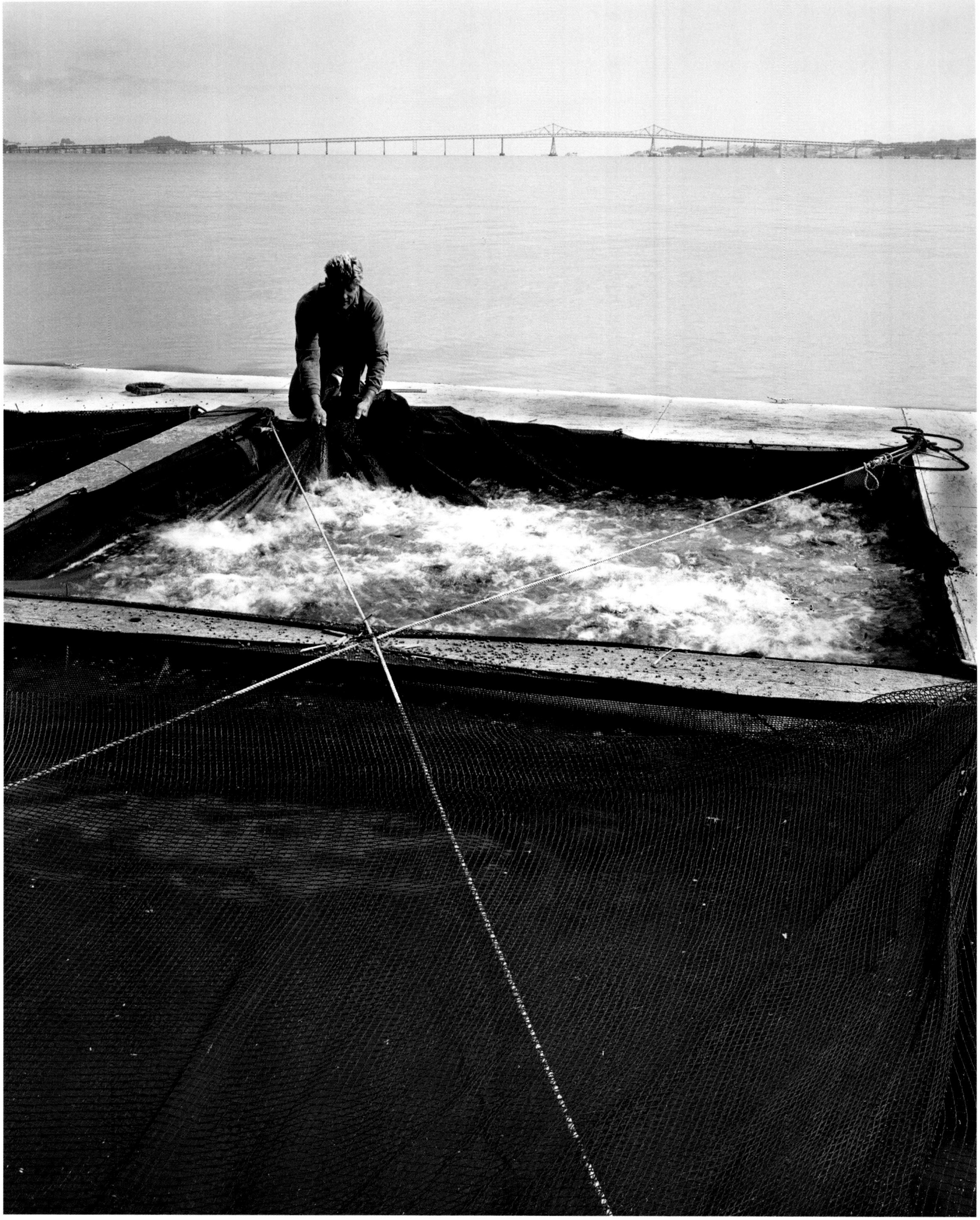

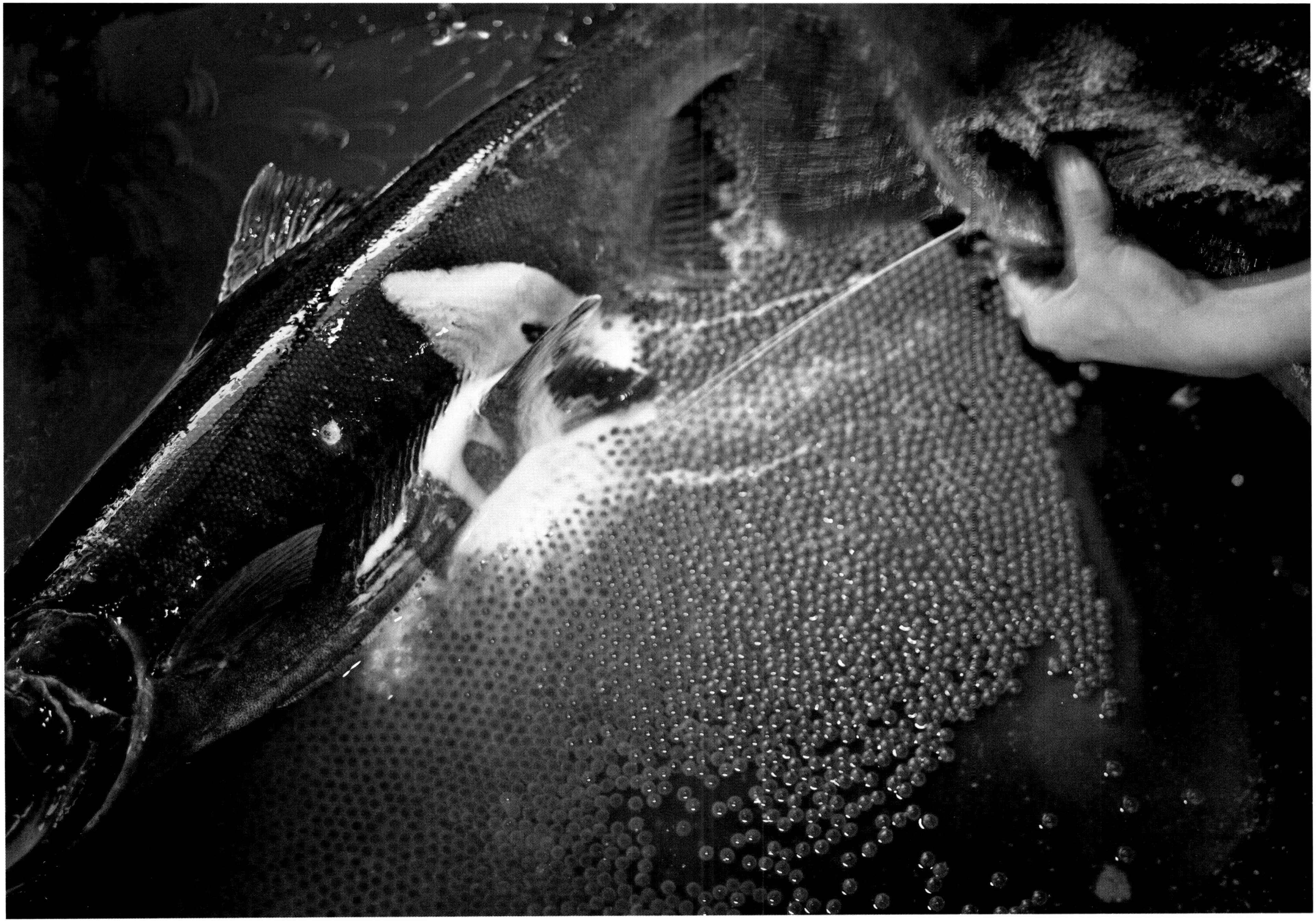

Milking salmon for eggs at Feather River Fish Hatchery, 1994

« Feeding time at a striper nursery, with Richmond Bridge in the background, 1997

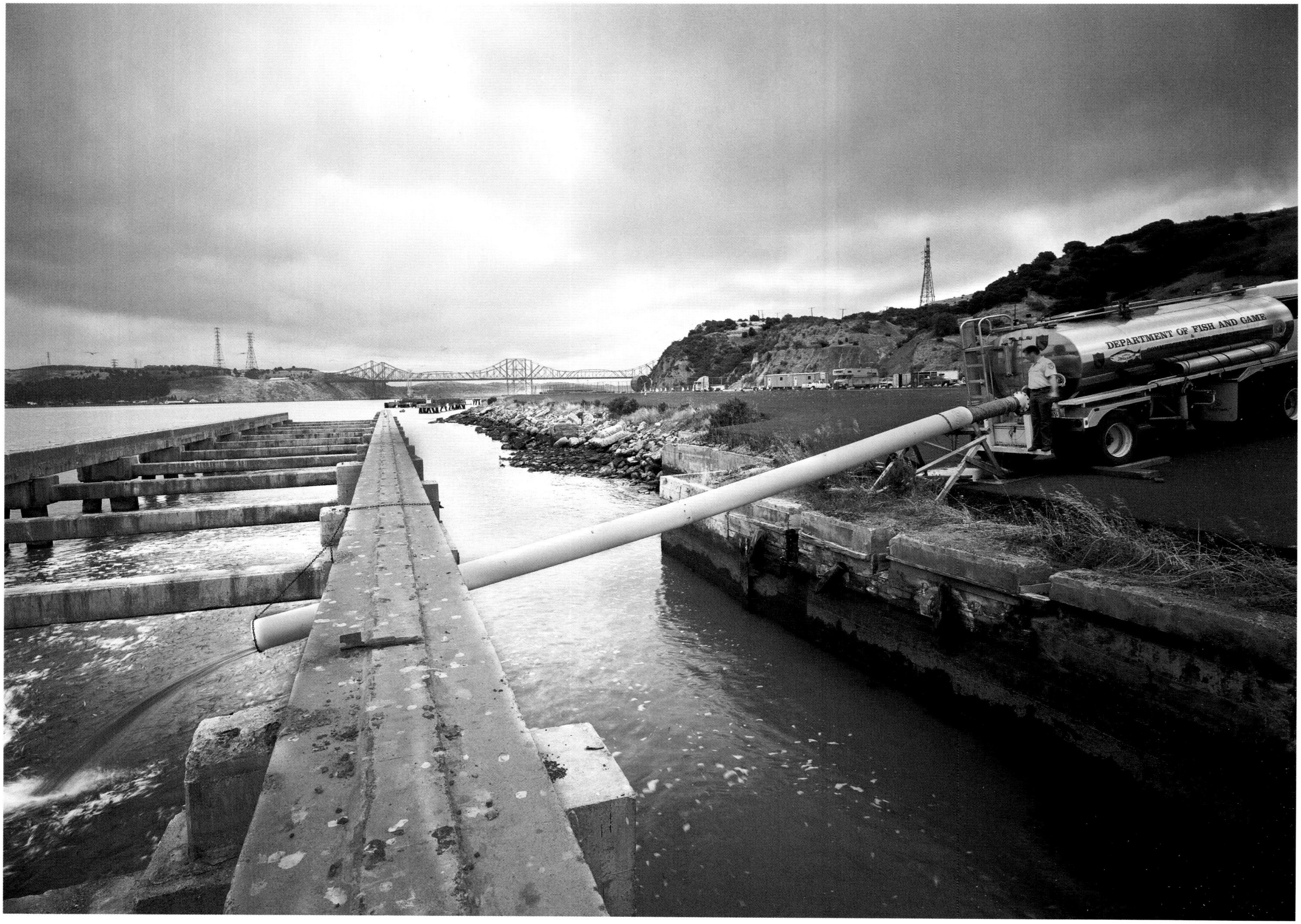

This page and facing page: Dumping fingerling salmon just west of Carquinez Bridge, 1997

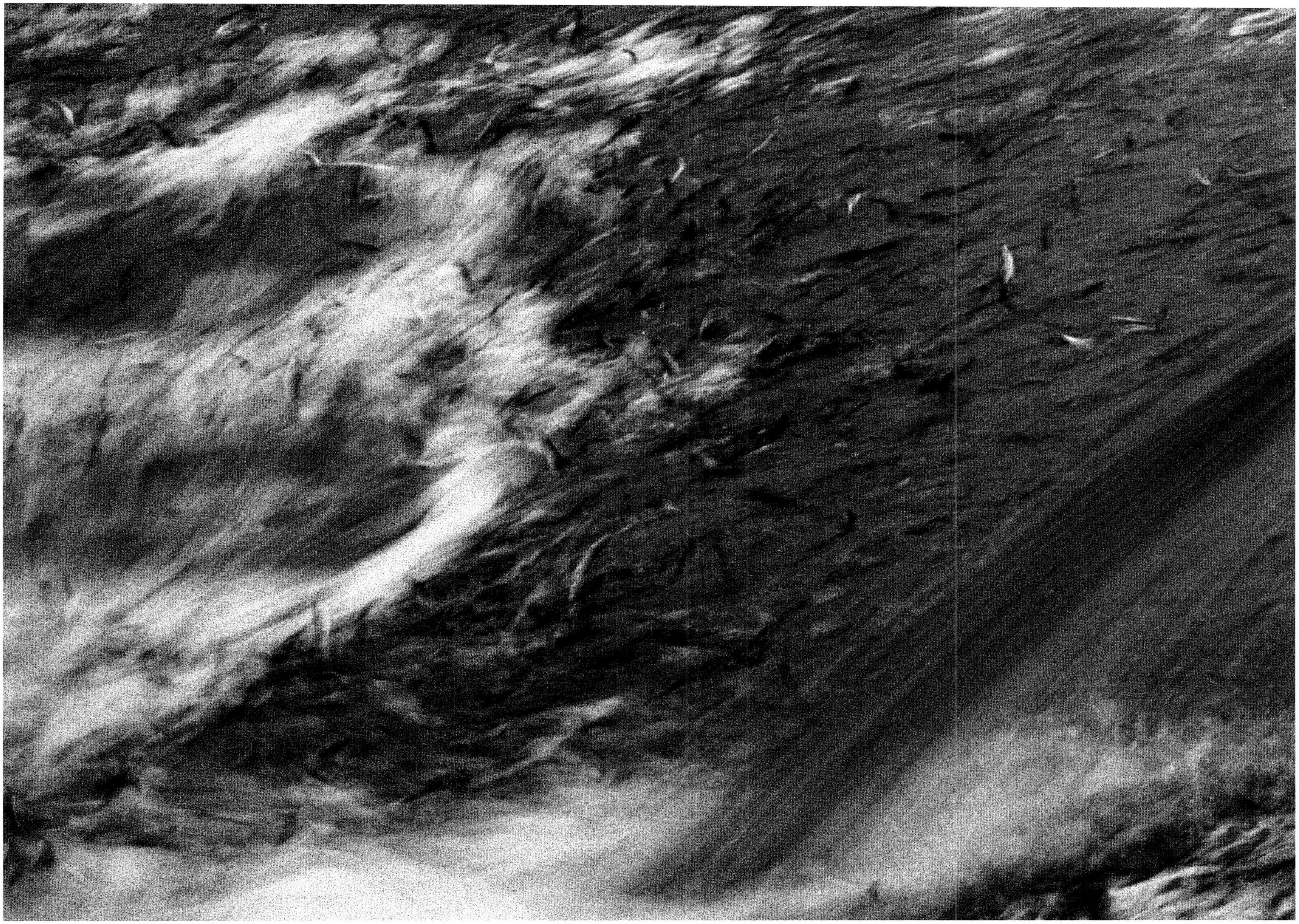

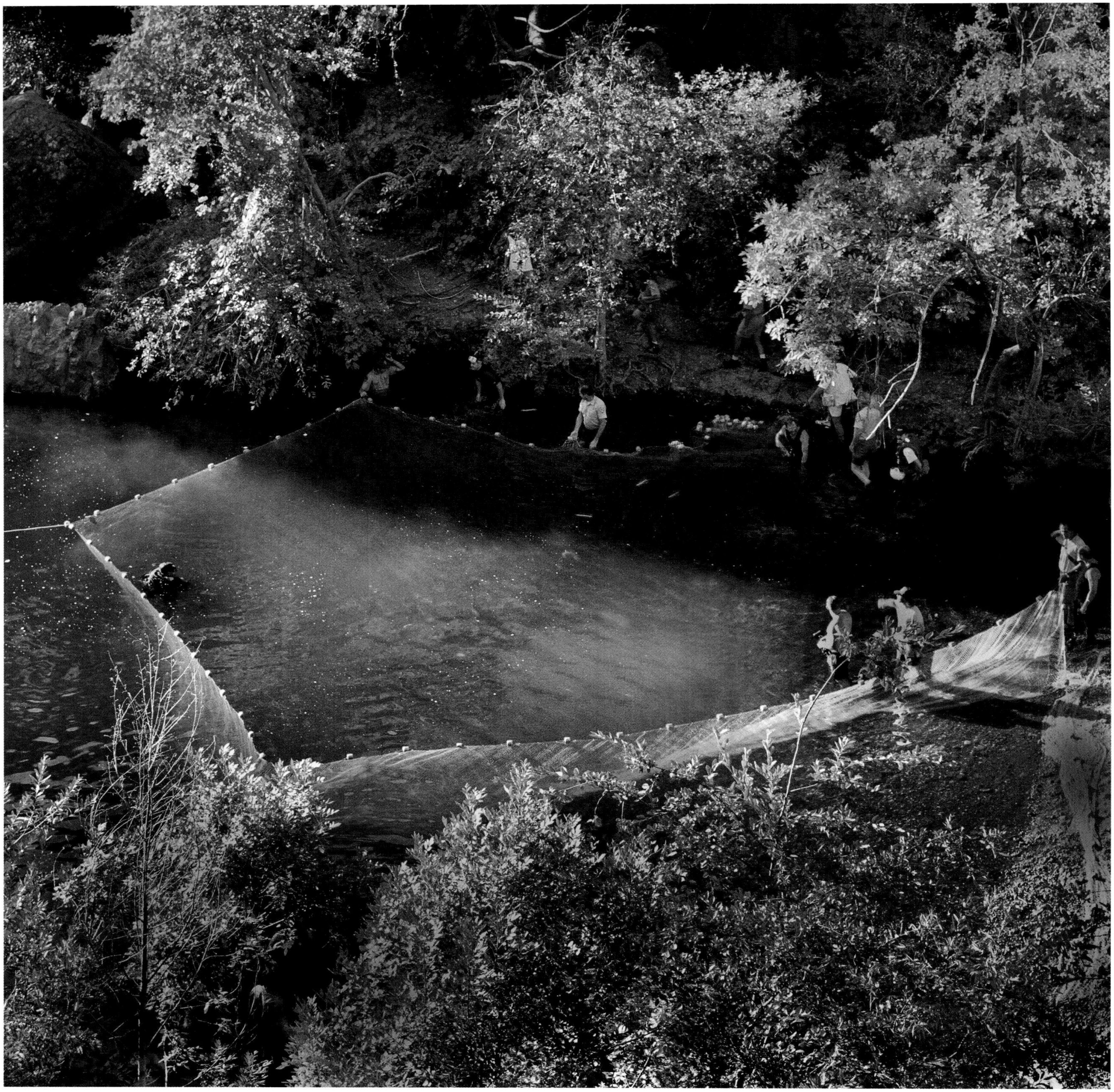

Salmon rescue on Chico Creek, 1996

Installation of an $80 million temperature-control device at U.S. Bureau of Reclamation's Shasta Dam, 1996 »

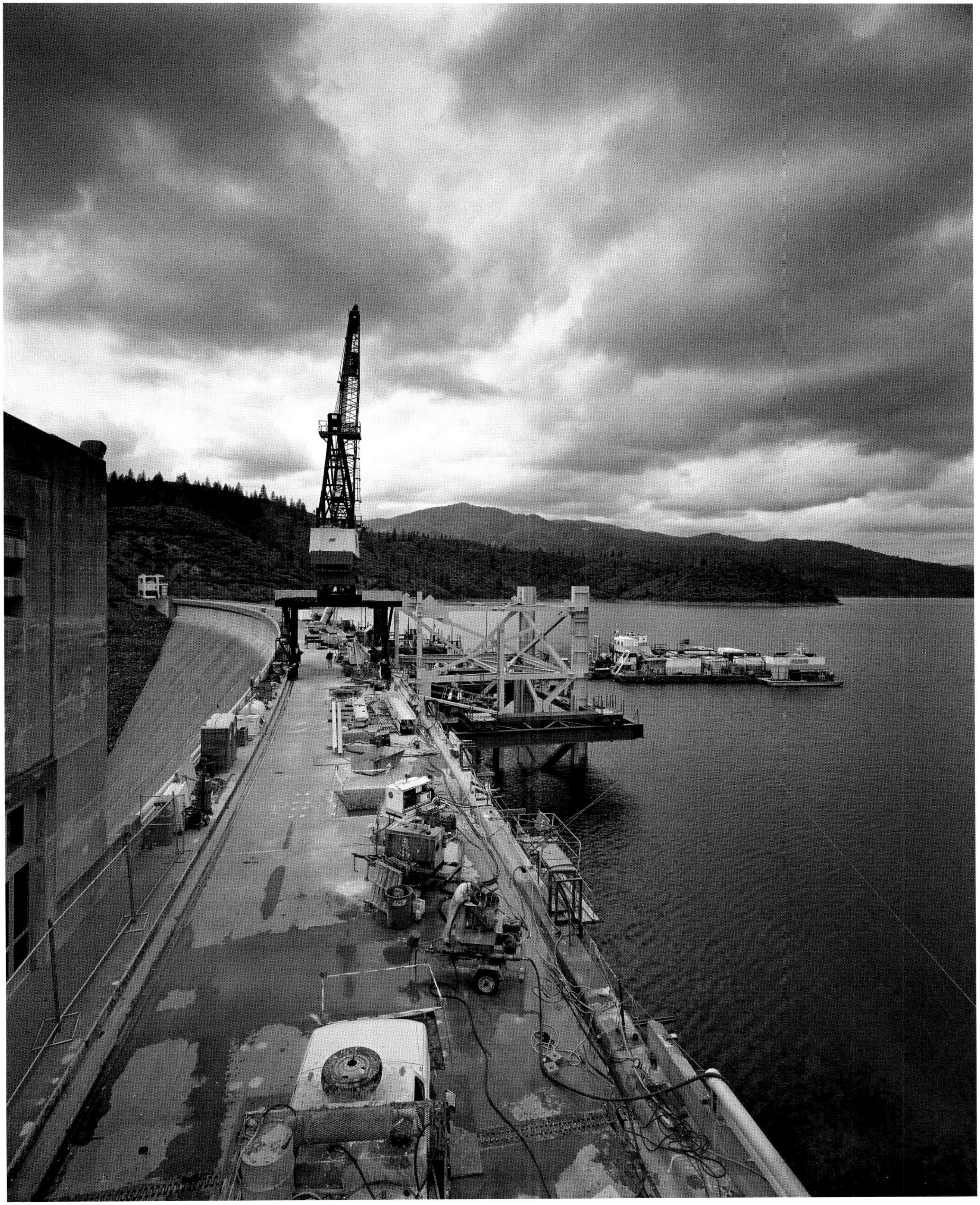

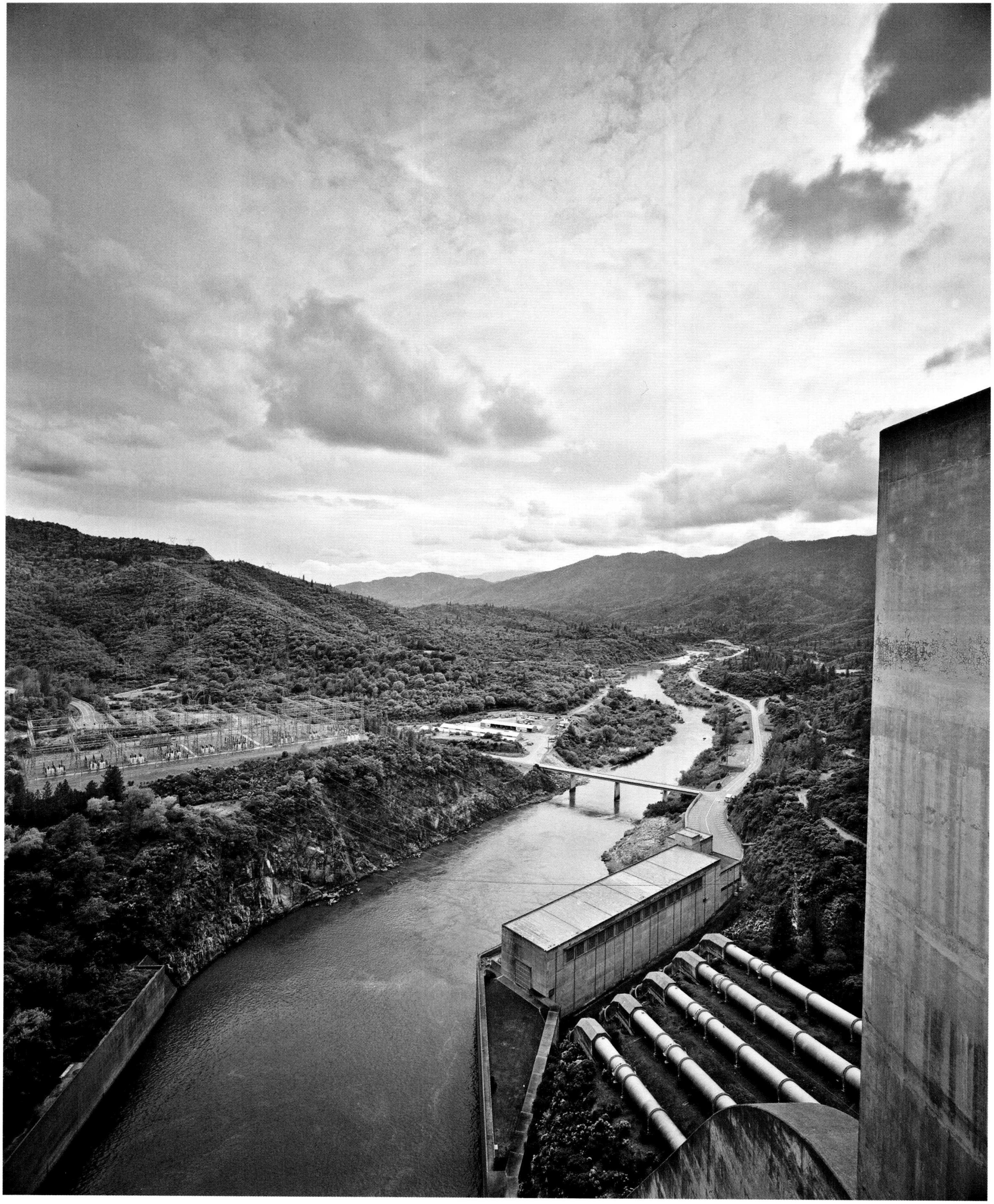

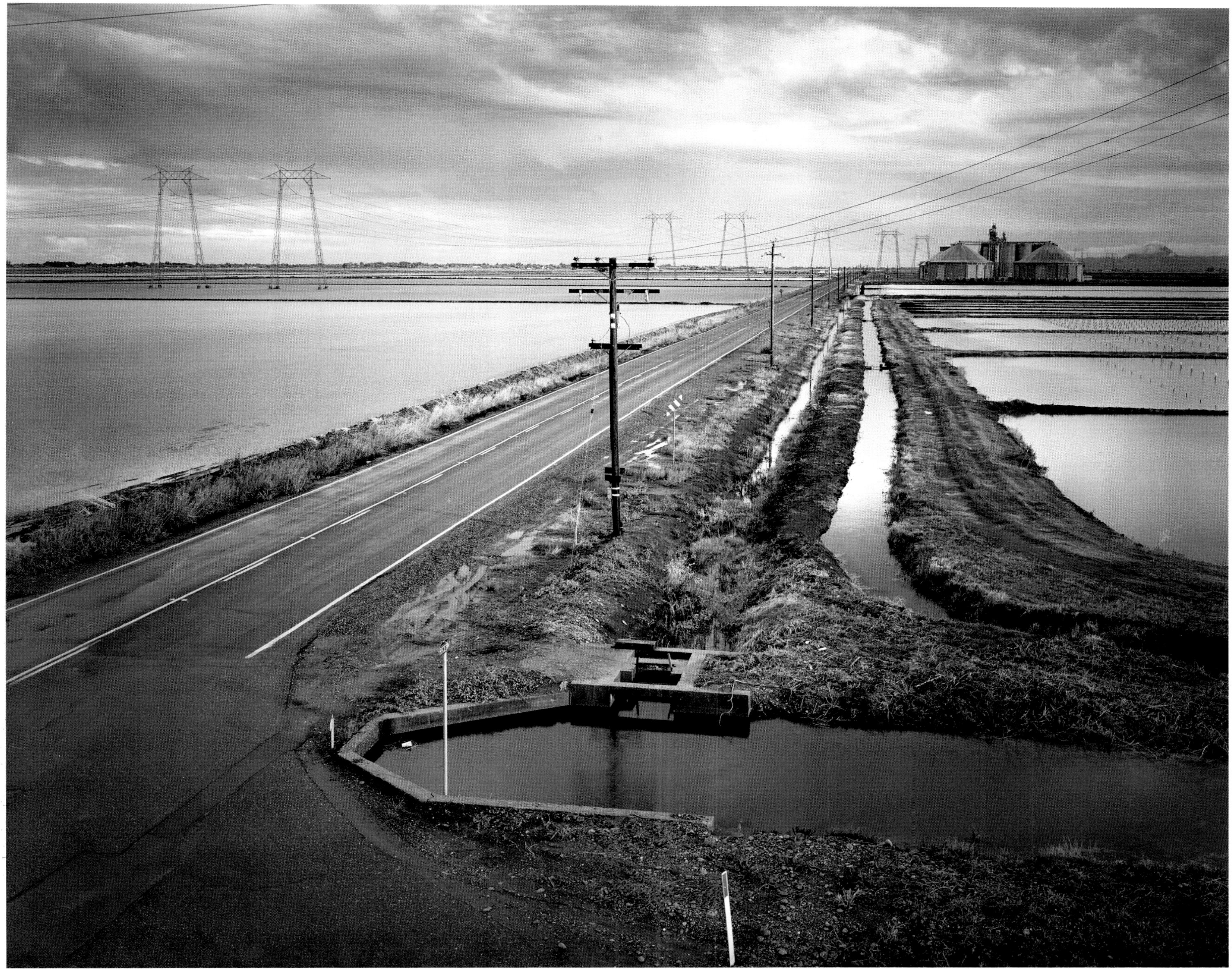

Rice fields adjacent to Riceton Highway flooded with water from California Department of Water Resources' Oroville Dam, 1996

« Looking downriver from U.S. Bureau of Reclamation's Shasta Dam, 1996

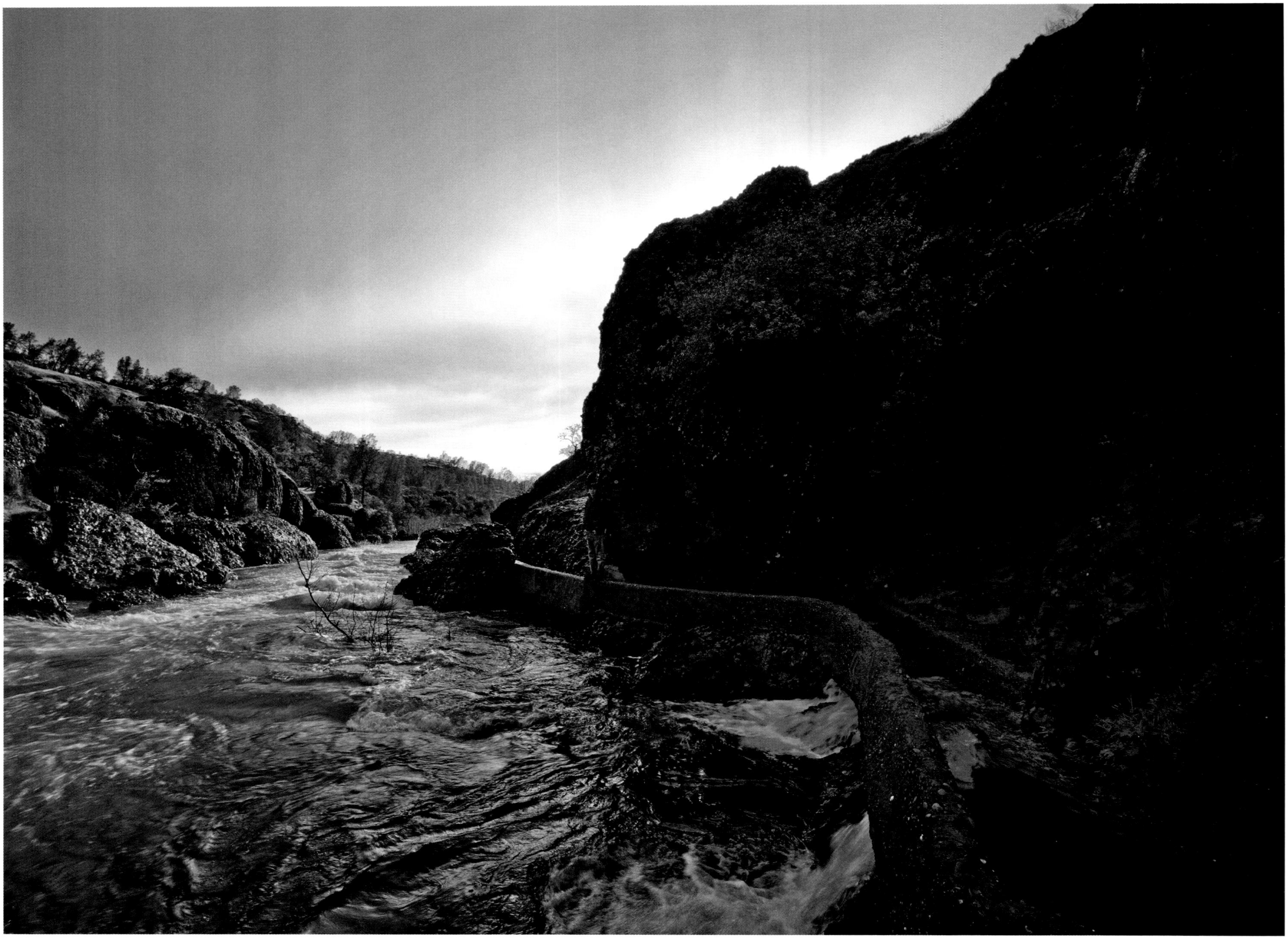

Leaky flume during high water, Chico Creek, 2011

C H A N G E

WATERSHED IS A GEOGRAPHICAL TERM defined by the topography of landforms and relation of waterbodies; it is also a word that means "crucial dividing point." On a mountain between two watersheds, we stand between separate worlds. We can choose to walk forward or go back the way we came. At a watershed point in our history, we can change, informed by evolved awareness, or we can continue to act without care, over and over.

Change a river, change the world. In the two hundred years since European contact with the Sac, we have transformed its renewing, flood-prone character to a modern system of engineered channels, dams, reservoirs, intakes, levees, bypasses, forebays, and afterbays designed to pump, divert, and store water. In doing so, we have pitted ourselves against water's inherent essence. We have assumed we can create more supply, although it cannot be done; we have put up hard barriers thinking water won't evade or erode their weaknesses; we have attempted to hold the river back from the sea. In countering nature, we've used every tool in our belt—digging, blasting, hauling, negotiating, stealing, publicizing, lobbying, marketing, dealing, diverting.

The sum of all our efforts on the Sac equals a system of monumental supply: today's waterworks for this single river serves over half of California's thirty-eight million residents, a demand that amounts to three trillion gallons (9 million acre-feet) a year for urban use and eleven trillion gallons (33 million acre-feet) a year for agriculture. Even for a river with an annual average runoff of about 7.2 trillion gallons (21.6 million acre-feet), a flow that surpasses that of every Pacific Coast river in North America save the Columbia, meeting such staggering demand is a tall order. California's population is expected to reach fifty million by 2020. When we achieve that number, the difference between water demand and supply is predicted to be between 25 to 75 percent of today's total urban water use, or 2.25 trillion gallons (6.75 million acre-feet). Today, even with increased efficiency in water applications, agriculture and related manufacturing requires over 80 percent of all water consumption. Landscaping is responsible for over 80 percent of urban usage. Having overtapped natural systems to meet demand in these two sectors, we have already failed to earmark sufficient water for the third but original user: the river ecosystem.

We have created a greatly compromised Sac: reduced streamflows, declined fisheries, increased water temperatures, floodplains walled off from recurring overbank flow that replenishes soil. Off-channel nursery habitat is no longer accessible to salmon, streams are prevented from meandering and therefore cut deeper, and invading, exotic plants and animals crowd out beneficial native species. Mounting hydraulic pressure from increasing peak flows sends waves of change through the Sac watershed. Downstream, 80 percent of the delta has been diked and drained. Dewatered land has subsided below sea level, and its levees are already weak and degraded by erosion and burrowing rodents. The delta is immediately in danger of permanent flooding.

Over all other environmental threats hovers the global game changer: warming climate. That it is occurring is not in question. That it threatens the future of the Sac and all rivers is certain. Snowpack quantity in the Sierra Nevada has decreased by 10 percent in the last century and is still declining; snowmelt runoff timing is shifting due to late-season increases in storm peaks; coastal groundwater supplies are threatened by saltwater intrusion from a sea-level rise on the California coast of about seven inches over the last century; droughts are becoming more severe and frequent, requiring more and longer irrigation needs for agriculture. Human requirements will intensify while State Water Project and Central Valley Project supplies remain subject to flow restrictions and limitations to protect threatened and endangered species.

The stage is set for frequent and growing unnatural disasters: mounting temperatures with loss of cool riparian habitat, longer droughts, and not enough of our lifeblood—*drinking water*—to go around.

Even without changing climate, water supply in California cannot meet demand for urban, agricultural, and ecosystem needs given our current patterns of use. Aquifers pumped too quickly cannot recharge sufficiently. Some are contaminated already by metals, salt, or nutrients and cannot be used for drinking water. In 2009, the difference between pumped groundwater and annual recharge amount (the deficit) was estimated to be 652 million gallons (1,956 acre-feet). Necessary for emergency stores, the groundwater is being used now—in much the same way we tap fossil fuels critical for emergency and medical applications—with a flagrant lack of conservation. No one is cranking down the faucet fast enough; we are all overspending critical supplies.

Some plans to meet increasing demand in California think well inside the box, falling back on the tired old call for *more:* more river withdrawals, more degradation of critical environmental protections, more pumping of fossil groundwater, more sales of publicly subsidized surface water to urban areas and desert farms, more infrastructure. One of the biggest and potentially most impactful plans for the living Sacramento is an undying proposal to revive the peripheral canal, a set of giant pipes that would divert south an estimated 1.3 to 2 trillion gallons (4 to 6 million acre-feet) per year of river water to serve a politically powerful contingent of users. Running east of the delta, further altering natural streamflows, the canal would hasten the demise of aquatic populations and unbalance the delta's freshwater-saltwater interface. Such short-term solutions divide all Californians along well-established battle lines, and they only put a finger in the dike of our need. Every time we fail to address demand and simply increase supply, demand rises to new heights. Worse, every time we increase supply in the old ways, we overtap the life-giving river.

Other plans bring fresh ideas to an old debate.

Research more than a decade old shows that existing technologies could be applied immediately to decrease urban water use 46 percent. Replanting drought-tolerant native species, if promoted without delay, could be extremely effective in relegating landscaping to lower-tier usage. Use of reclaimed water could increase fivefold fairly quickly; similarly, industrial use could become more efficient by 20 percent—today—though the institution of currently available water-saving practices. Coordinated shifts in agricultural cropping patterns could lower annual net water demand by California farmers some 1.14 trillion gallons (3.42 million acre-feet) while increasing farm income immediately. More than 652 million gallons (1,956 acre-feet) per year could be shifted to environmental needs from both agricultural and urban uses, using new management techniques that benefit all three categories of demand.

Not surprisingly, one strategy to protect water supply as well as health of human populations is to preserve intact ecosystems. One of the most important, by all accounts, is a river with sufficient instream flows and a diverse habitat corridor. The Sac, with its forest enclaves and species strongholds, offers refuge that fortifies life on earth: survival of species requires core reserves, connected habitat, and preservation of critical species. The Sac is one of California's last major core, connected, and critical ecosystems.

Along the Sac, nonprofit and community groups like the Sacramento River Preservation Trust have made this connection: they organize tours and float trips to educate the public about the importance of the river. Private residents and business partners pioneer water harvesting and water conservation technology. Government agencies such as the California Department of Water Resources, California Department of Fish and Game, and U.S. Fish and Wildlife Service support and oversee projects that integrate science, habitat restoration, and management of flooding and fisheries. Community is stirring.

Through restoration of diverse native forests on streambanks along the Sac, community-agency partnerships are already bringing back declining species of birds and fish. Restoration of salmon passage through dam removal on important tributaries to the Sac, such as Butte Creek outside Chico, prevented loss of a Chinook salmon run critical to survival of the species. A collaborative, incentive-based anadromous fish screen program has slowed the unintended, massive take of Chinook salmon, steelhead trout, and green and white sturgeon into water diversions throughout the watershed. Planting native vegetation in riparian forest assemblages, such as those in River Partners' more than sixty restoration projects ranging in size from two to two hundred acres on the Sac and tributaries, preserves natural stream processes, cools the river, and reduces greenhouse gas emissions. We thereby slow climate change and restore streamflows. Projects that conserve groundwater protect drinking-water stores; many of the same projects reduce the risk of flooding, thereby preserving communities. Still to be seriously addressed are plans for population equilibrium—the true way to control demand, the one way to stop species extinctions. The native habitats on which our lives depend are resilient only when we don't exceed the carrying capacity of the planet we call home.

State and federal governments, once joined in a mission to impound rivers, are now collaborating to find solutions to water shortages and flooding. The CALFED Bay-Delta Program (now the Delta Stewardship Council), begun in 1994 and joining the strengths of federal and state agencies, made plans in 2000 to renew its approach to river systems. The plan lays out steps for preserving water quality, conserving and recycling water, reestablishing ecosystem health, studying and improving levee systems, and storing surface water and groundwater. Underlying the plan is a vow to base all change on watershed science. With the understanding that it will take thirty years and at least $8.7 billion to implement, the plan promotes cooperative restoration.

Fish and wildlife are to be protected by water purchases dedicated to ecosystem flows; 114 billion gallons (342,000 acre-feet) annually have been assigned as instream water for aquatic species, without reducing flow to farms and municipalities. It's a beginning.

Today, for the Sac, there is no going back to life on an untamed floodplain where conjoined freshwater marshes teemed with wildlife, human diversity, and aquatic life in unimaginable numbers. Many say there is no sense dreaming about the past. From the watershed divide we straddle today, though, we can choose to advance toward our best possible future. Those devoted to the Sac know that its well-being and ours—not only that of Californians but also of the world—are joined as surely as hydrogen and oxygen bond in water. From this particular divide, at this moment in history, the Sac and other wildly fertile and productive watersheds are pointing the way from ruin toward renewal.

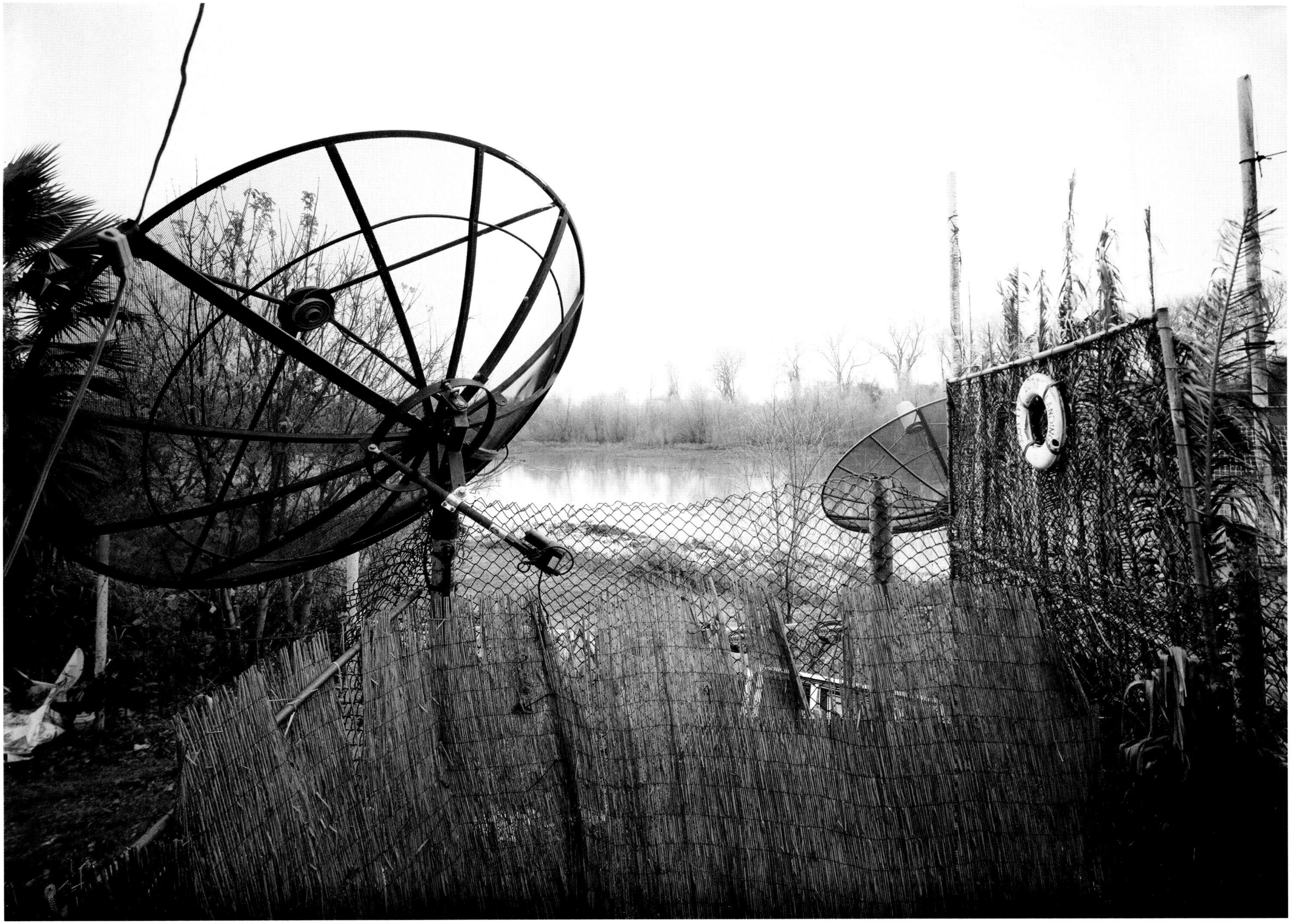

Communications at Scotty's Landing, 2005

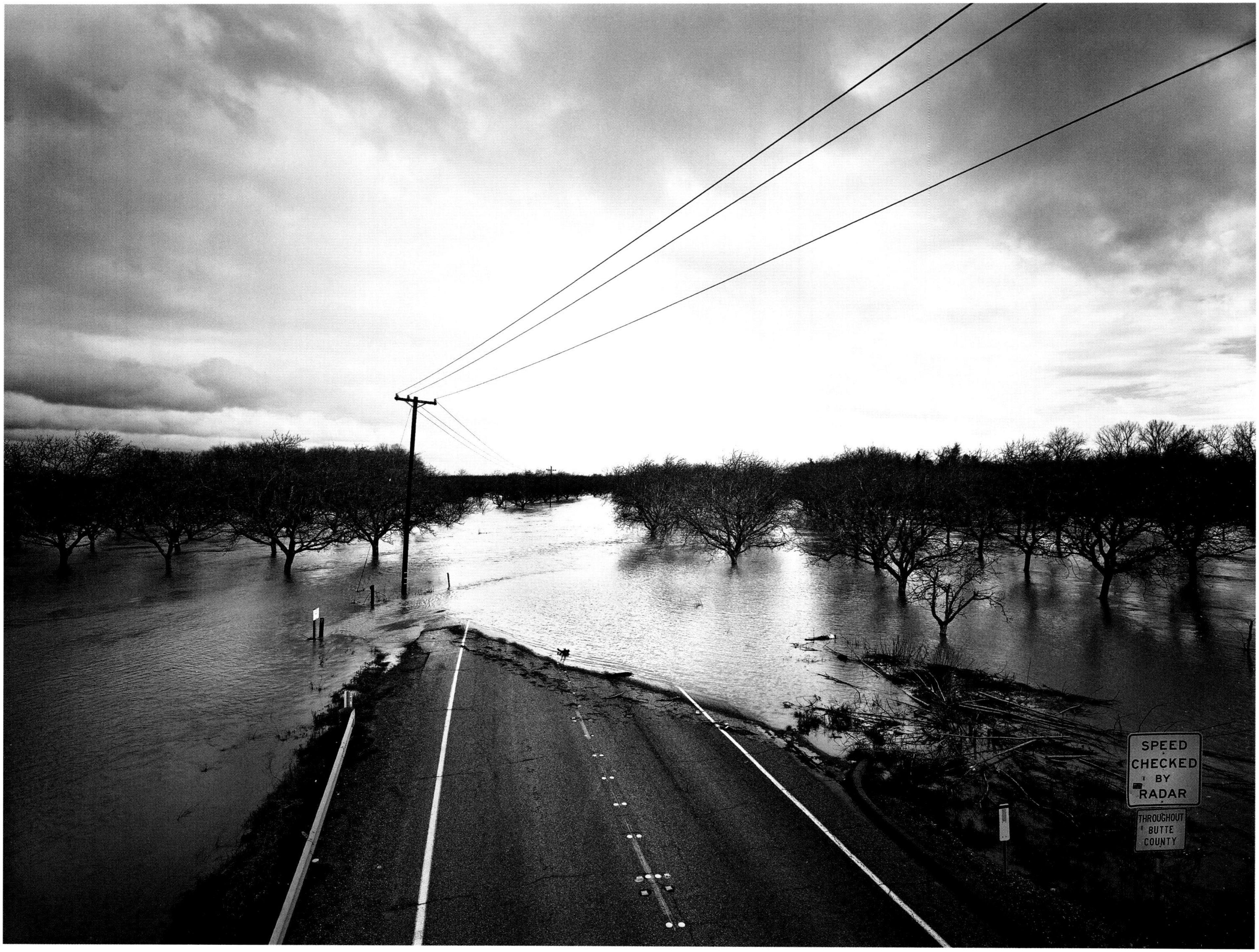

"Speed Checked by Radar," Ord Bend, 1997

Disabled parking space at Butte City Launch Facility, 1995 »

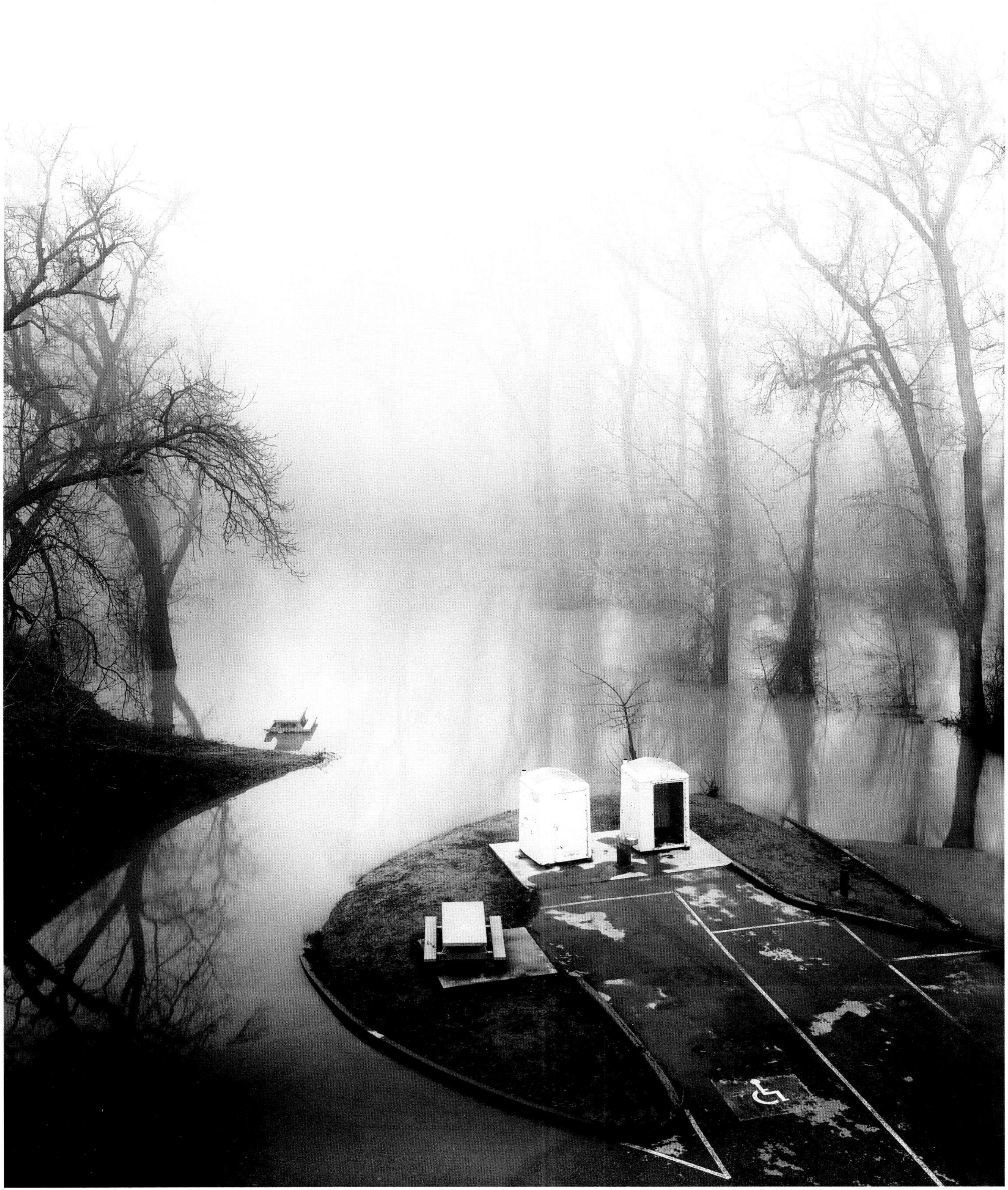

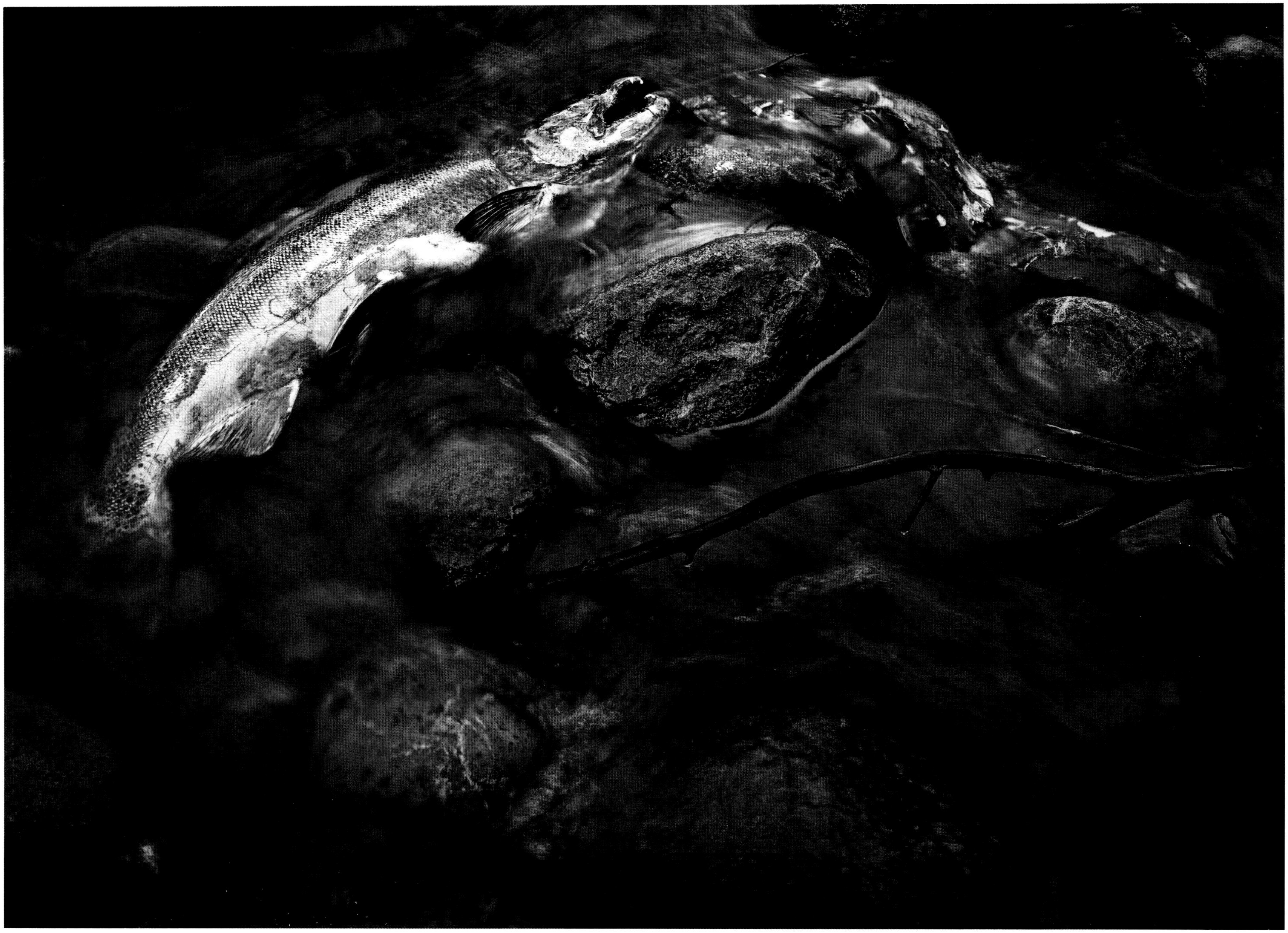

Two stages of a salmon's decay, 1996

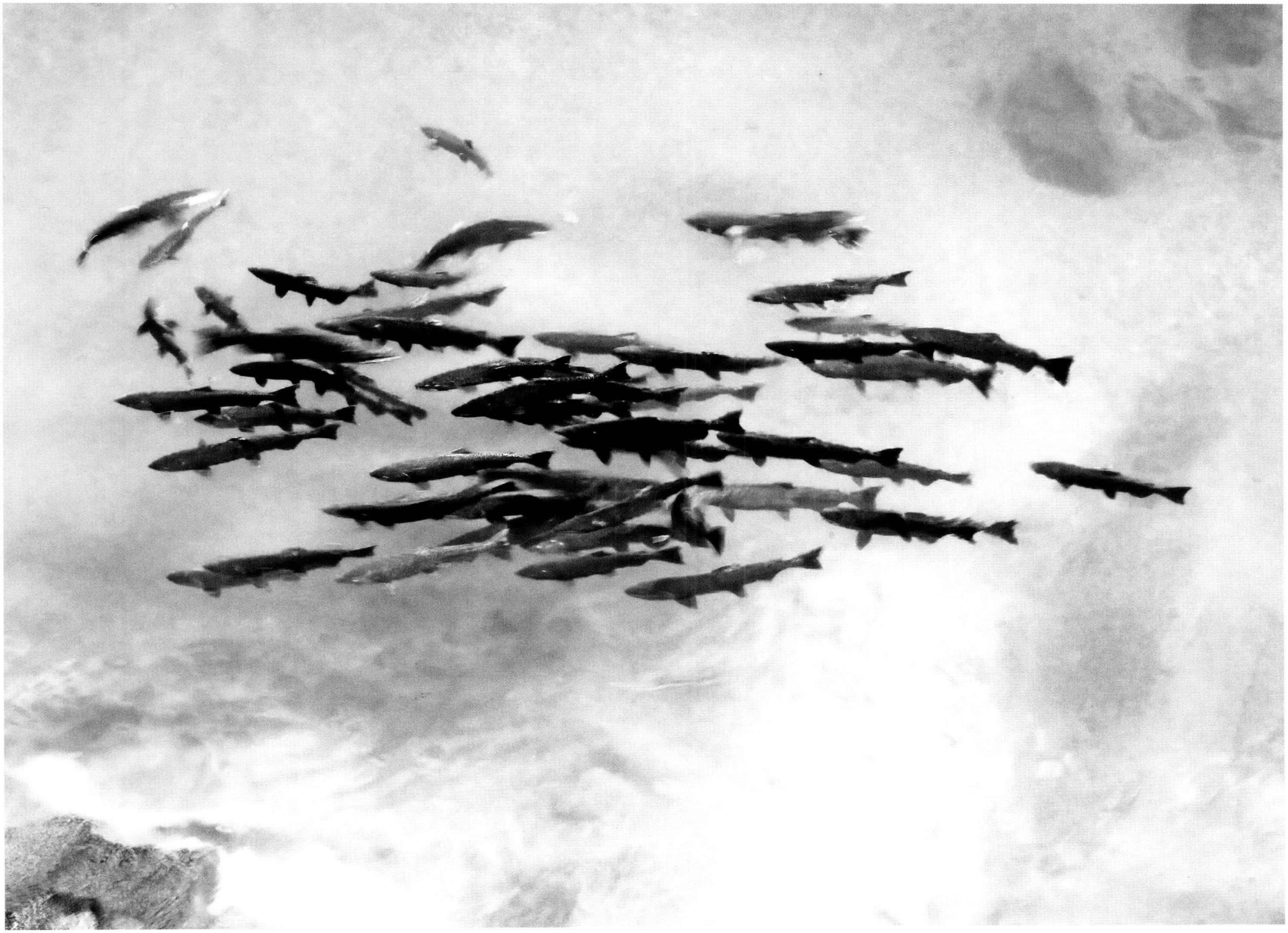

Salmon at Whiskey Flat, Butte Creek, 2001

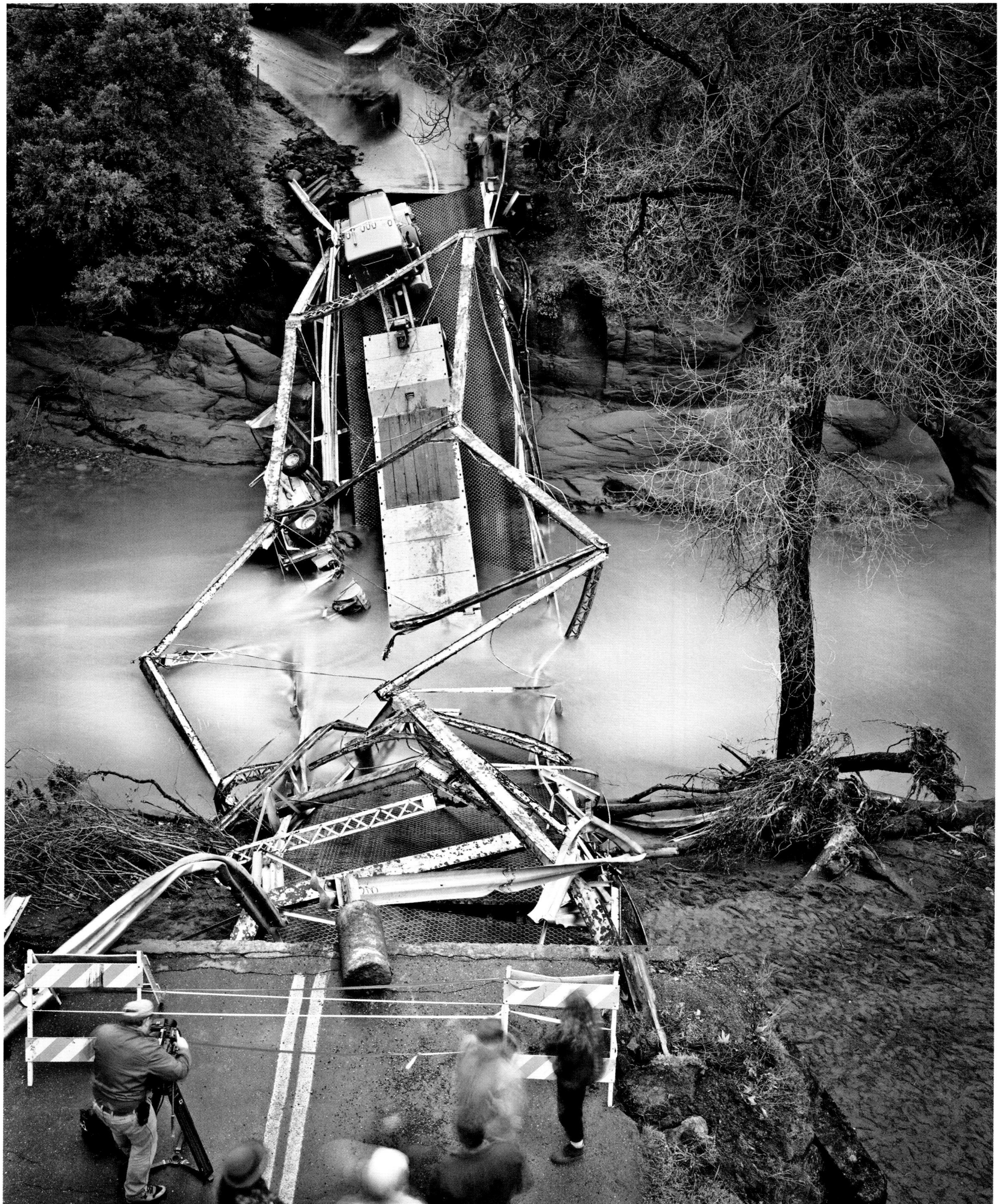

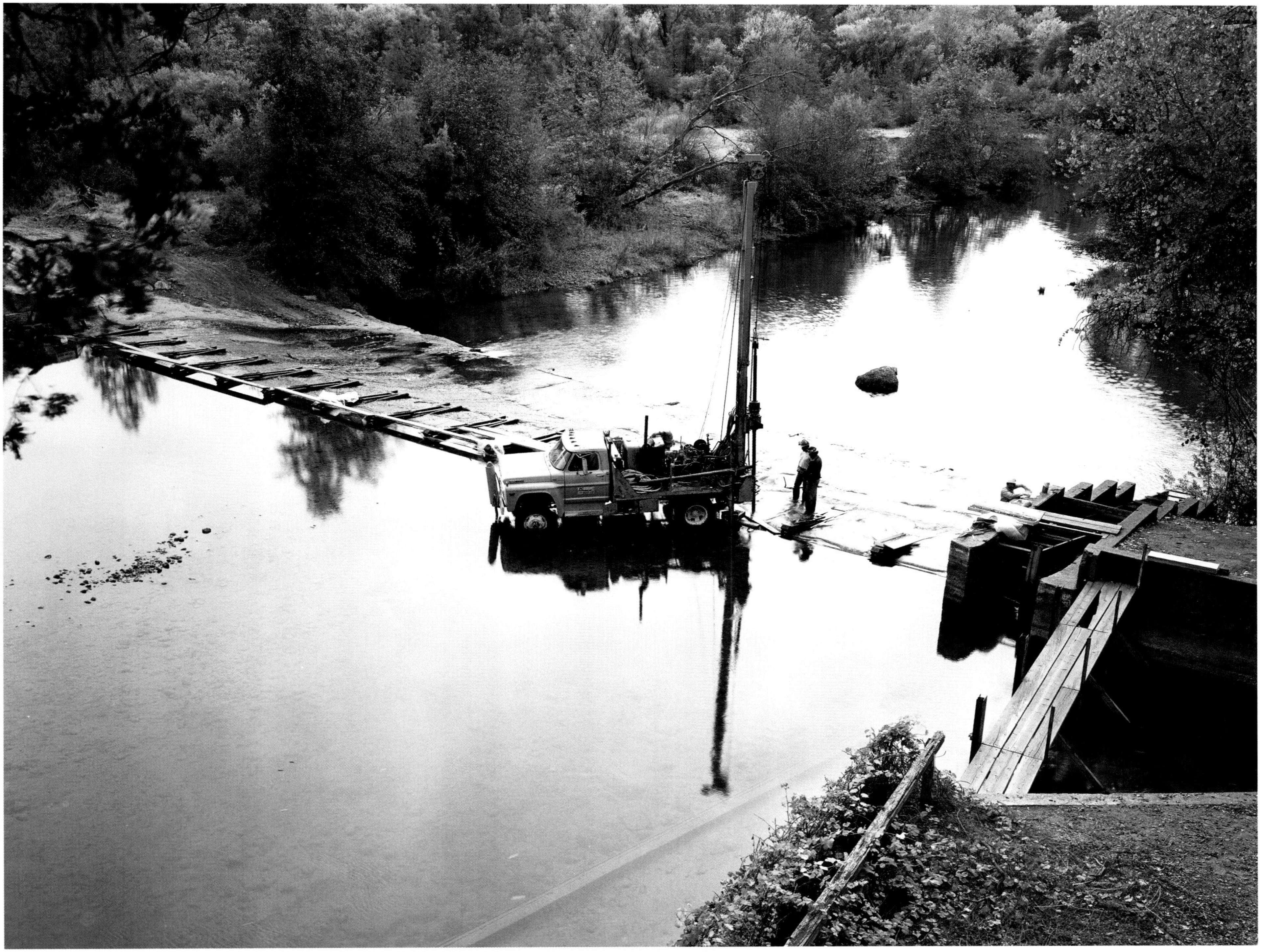

Geologist extracting a core sample at Parrott-Phelan Diversion Dam on Butte Creek, prior to reconstruction of fish ladder and installation of fish screen, 1995

« Butte County Bridge #17, Centerville, 1997

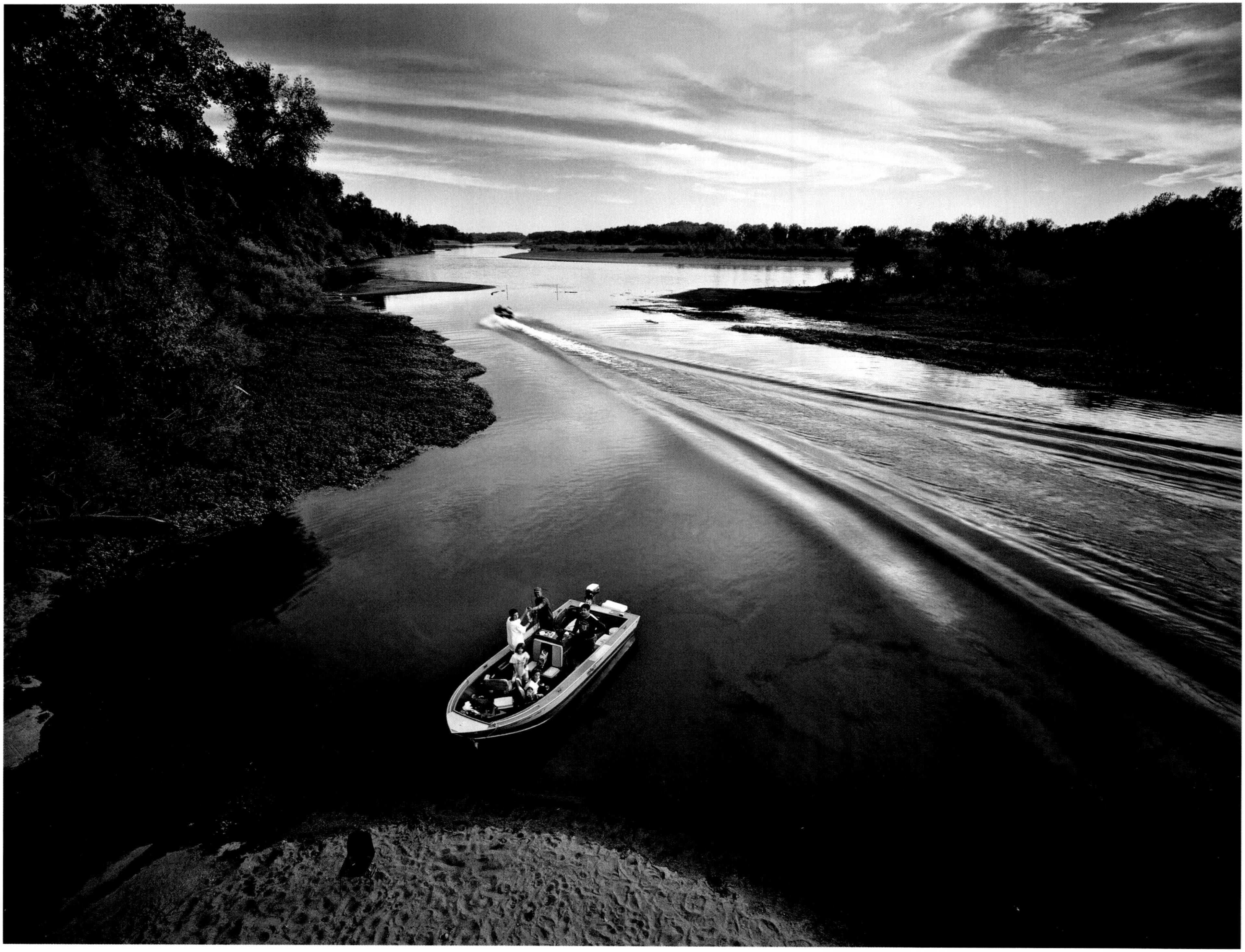

Fishermen at Chico Landing, 1996

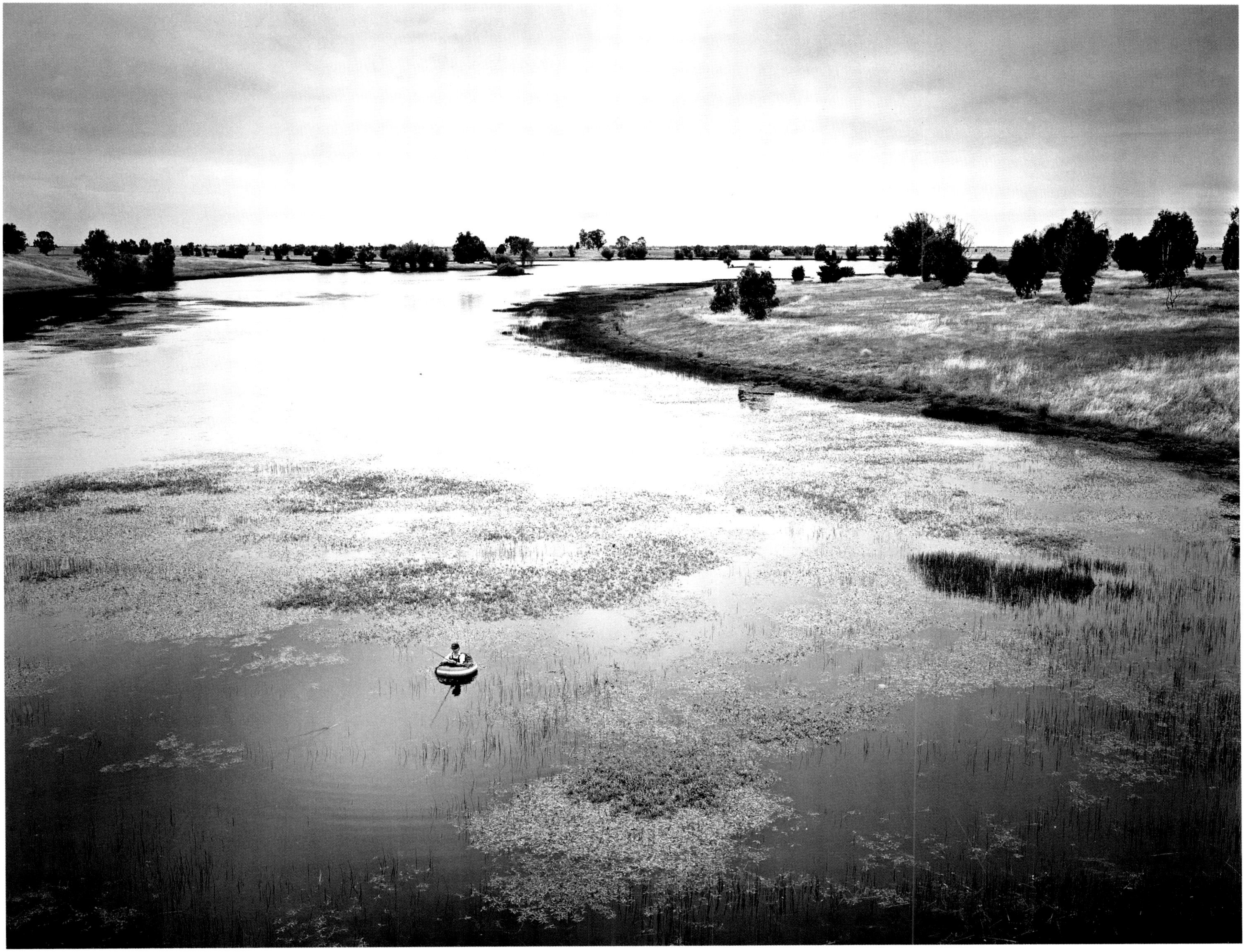

Lone tube fisherman at Blodgett Reservoir, Rancho Cordova, 1994

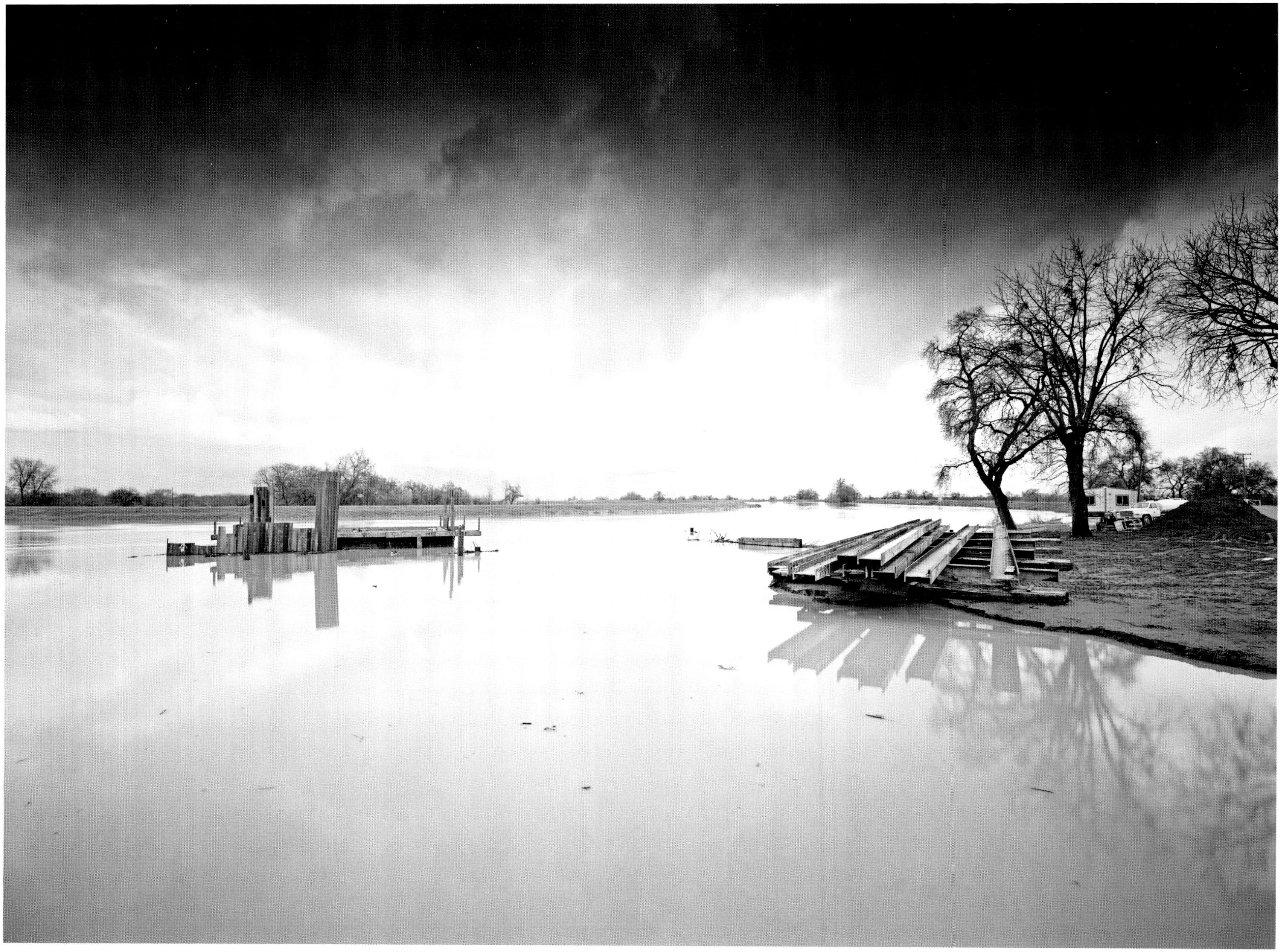

Construction of Wilkins Slough Positive Barrier Fish Screen, 1998

Flooded prune orchard, which has since been replanted with native grasses and trees by the Nature Conservancy, just west of the Butte City Bridge, 1997 »

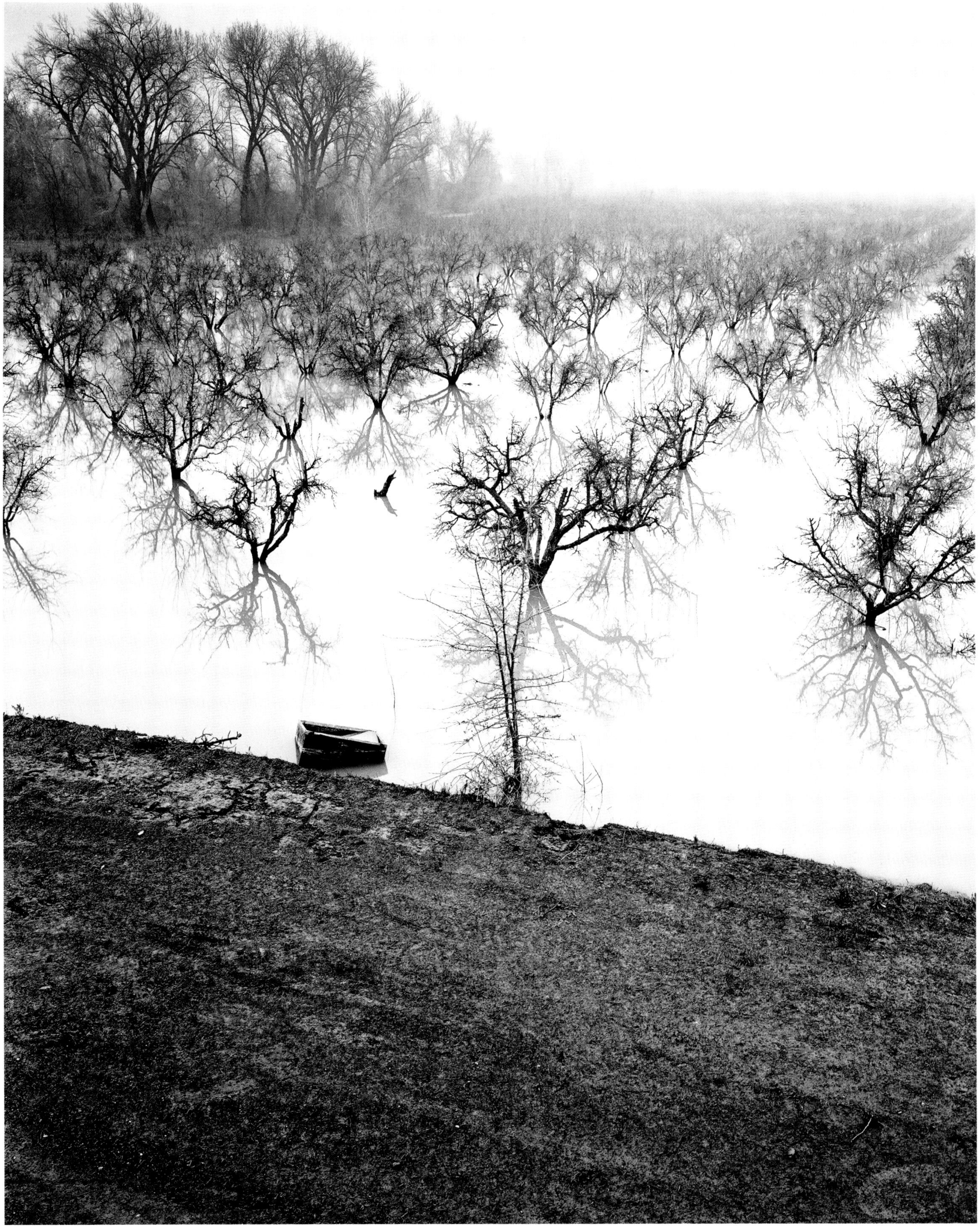

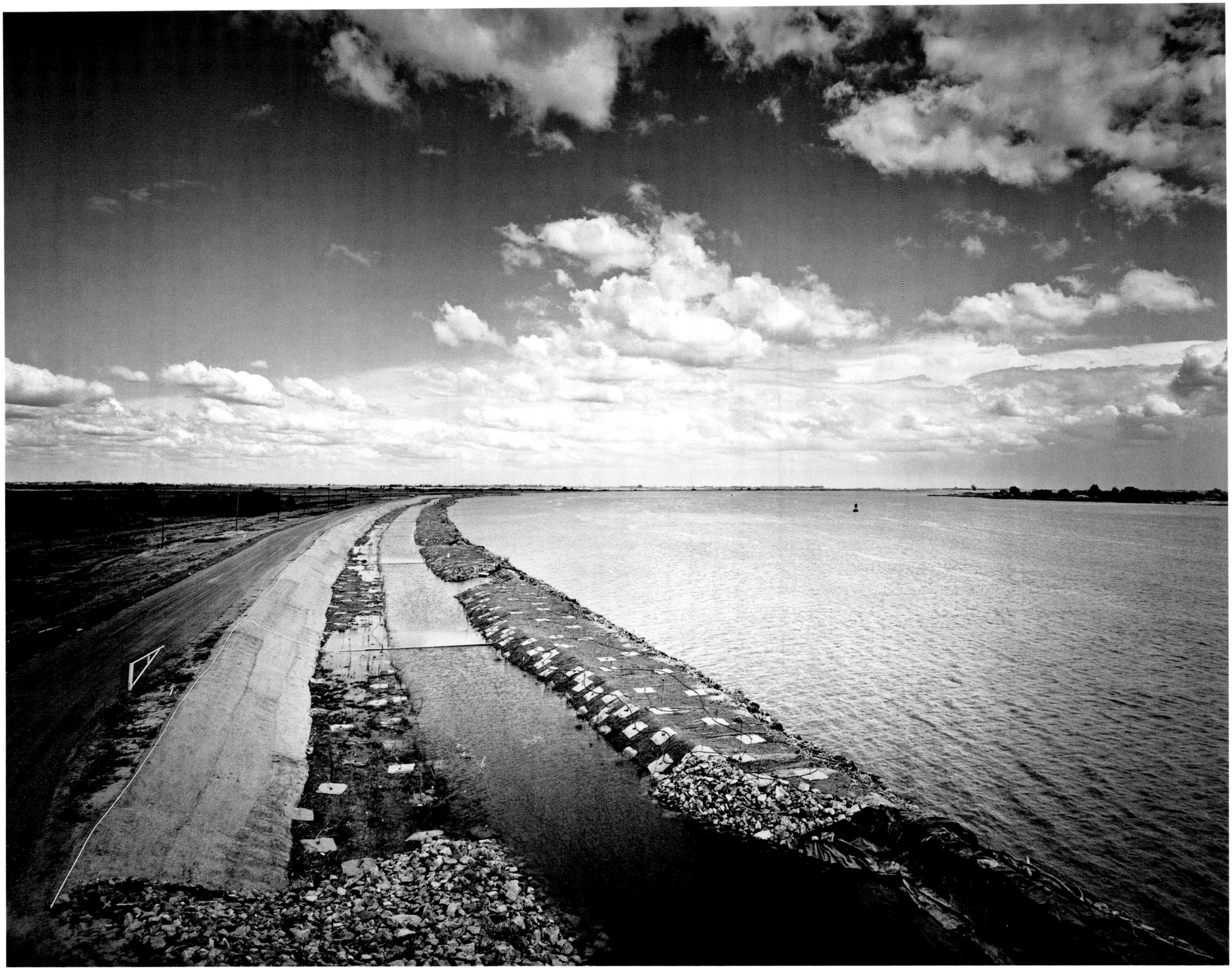

Revegetation of levee at Twitchell Island, Sacramento Delta, 1999

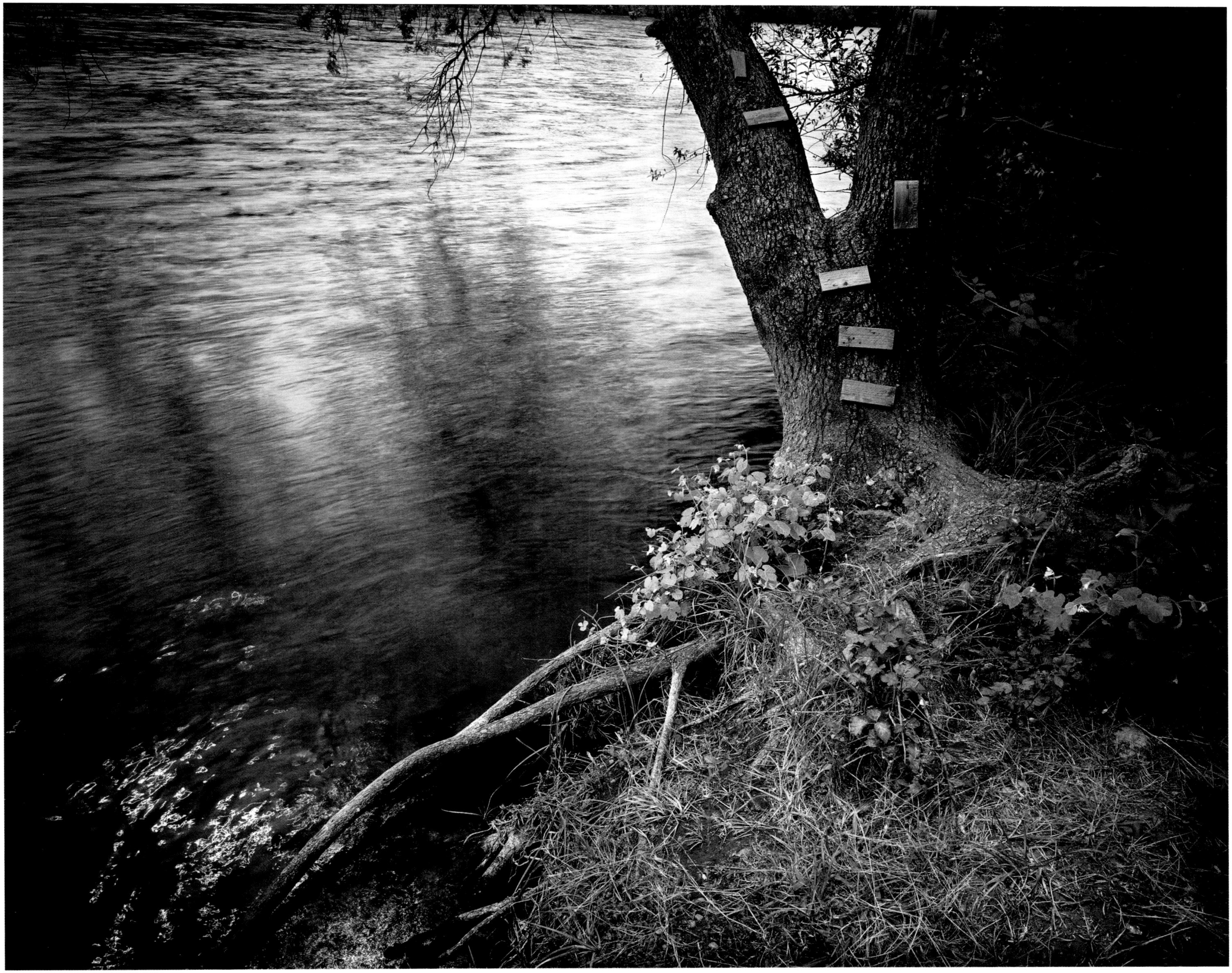

Steps to climb tree and jump into the Sacramento River, Redding, 2009

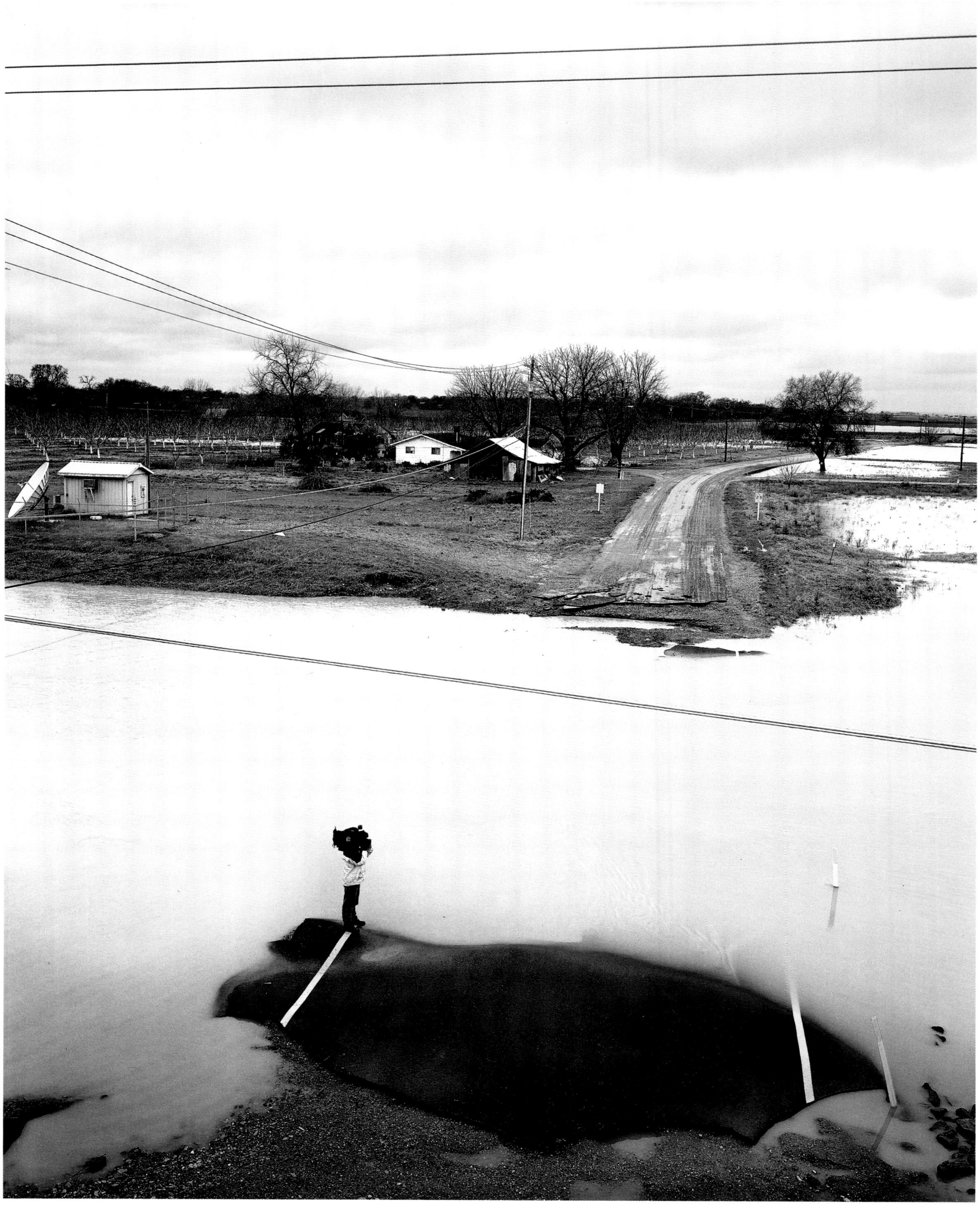

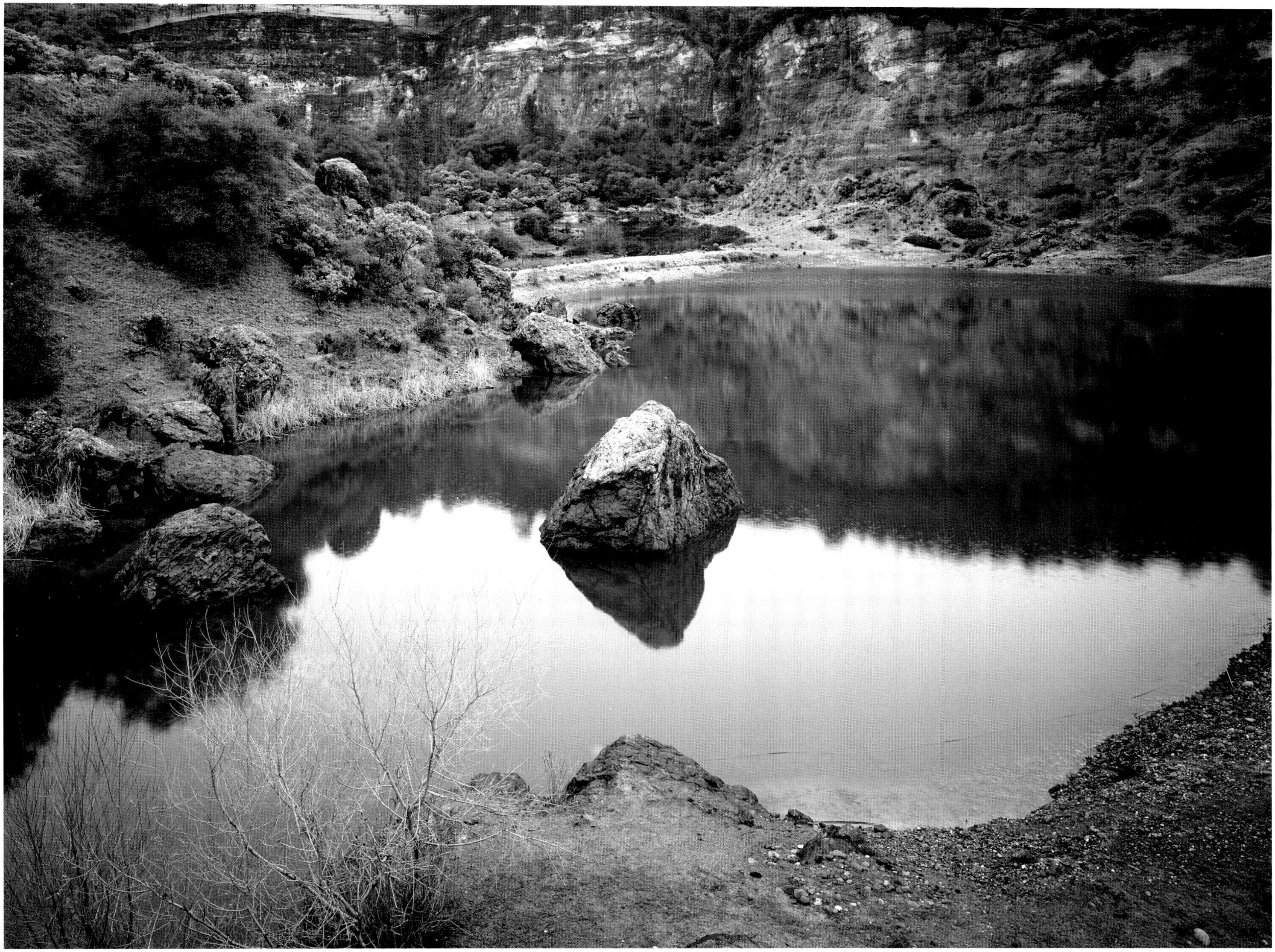

Blue Gravel Mine operations on the Yuba River, 1996

« News videographer at flood site in Tehama, 1997

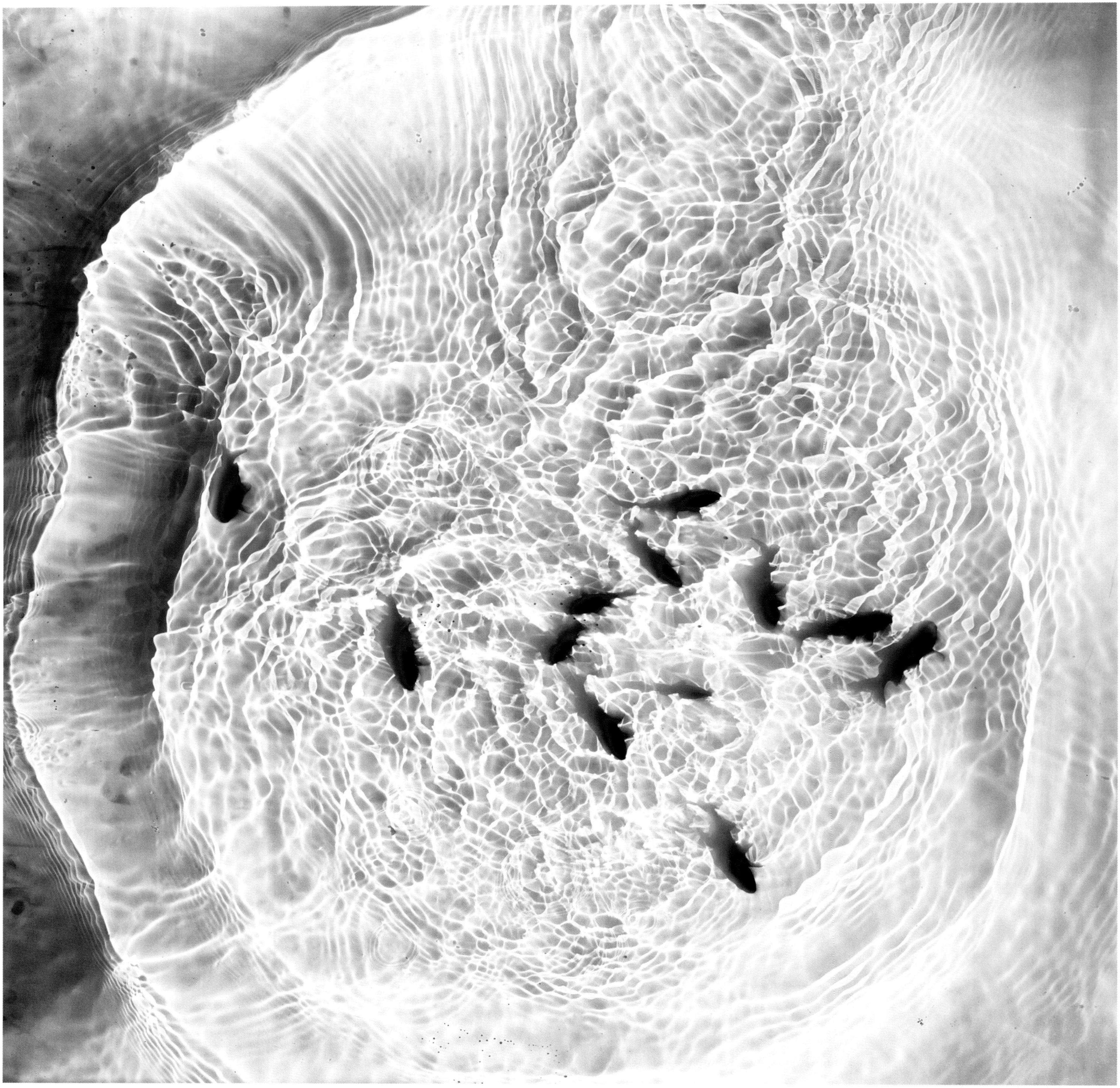

S A C R A M E N T

GHOST STRUCTURES REMAIN along the Sac, monuments to a time of ignorance about the scars of control. *Sacramento—river of devotion, allegiance, reparation.* We call the river by a name that confirms our awareness of its greatness. We haven't yet allowed its spirit to guide us to enlightened consciousness.

Water reverence in other cultures is offering clues.

For centuries, the enduring Hopi tribe of the American Southwest has used water as sparingly as sparrows at a bath, transferring rinse water among cups and bowls to lengthen the time of its utility. In part, this intentional application is due to water being scarce in their region; in part, such care is taken because water is respected and considered holy. Water reverence renders the people humble and blessed. On certain islands in the Pacific Ocean, where rising sea levels are ruining groundwater wells, running a faucet for more than a few seconds draws too deeply from stores of harvested rainwater. A profligate visitor, unused to taking such care, is gently reminded that the resource is finite. In some European alpine inns at high elevations, where water is difficult to attain, visitors are allowed no more than a few gallons a day for cleaning; metered tanks dole out the appropriate amount. Water awareness, rather than creating hardship, joins the awakened in a sacred journey. The journey is worth making for its own sake.

Save a river, save the world. Great minds may not always think alike, but when it comes to nature and rivers, evolved thinkers speak as a chorus. They tell us that our circle of compassion must embrace all living things. Those of us not naturally motivated to protect other species as we protect our own need only be reminded that we are linked by our common watercourses. As they thrive, so thrives the human race.

Meanwhile, the river speaks its own language of survival. Along the Sac are lovely, wild refugia whose richness connects uniquely with our souls. The green world stirs our hearts and our desire to be part of it, in the connected way of nature. Yuki wisdom says,

The rock did not come here by itself.
This tree does not stand here of itself.
There is one who made all this,
Who shows us everything.

If there is one entity in this world that creates generously and shows us everything, it is a wild river. If there is a river whose potential for biological richness and natural wealth can lead us to wholeness, it is the Sac.

El Rio del Sacramento. River of redemption. Miraculous river.

« Water as lens, 2006 (photogram)

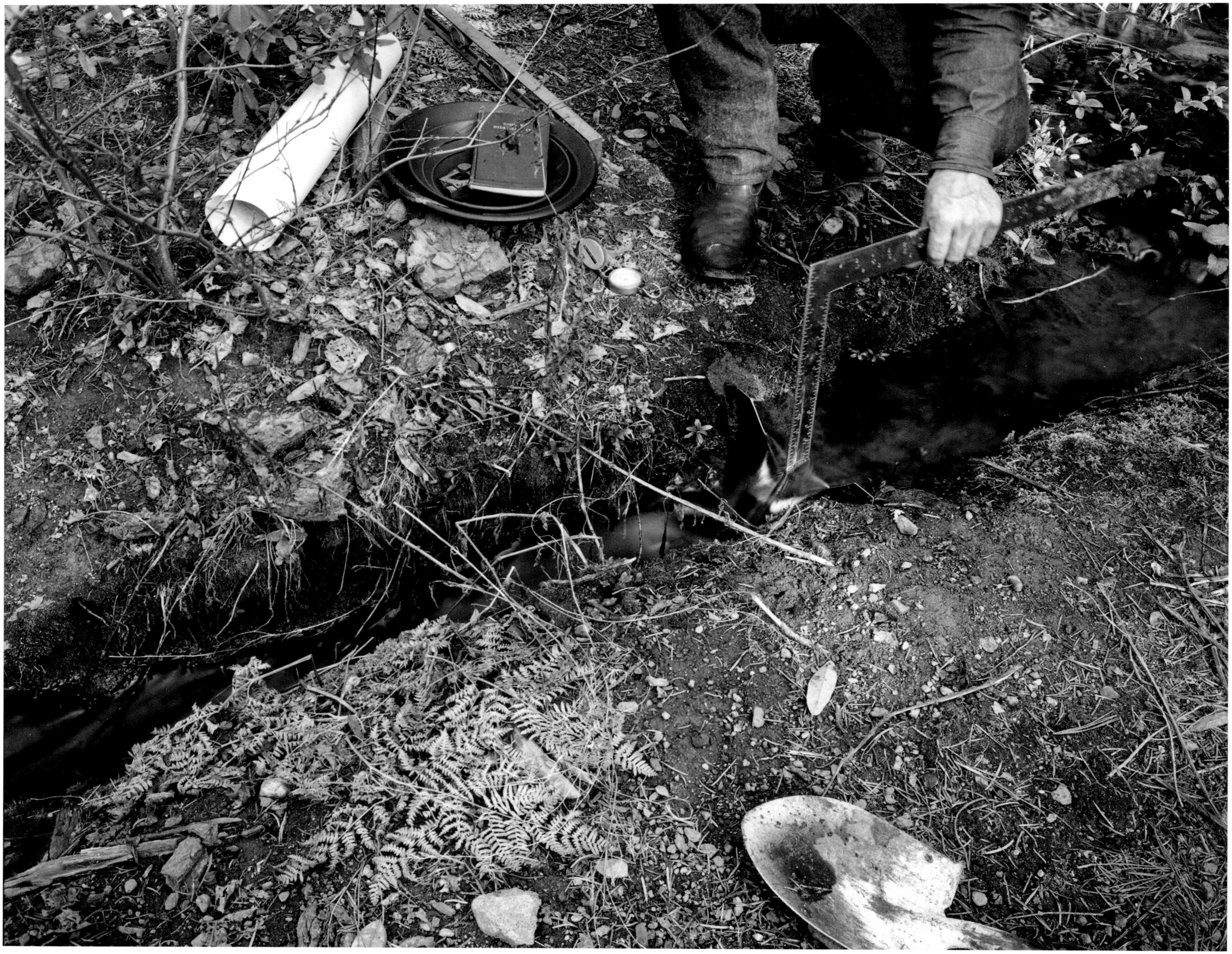

Three miner's inches, Dix Mine, Butte Creek, 1994

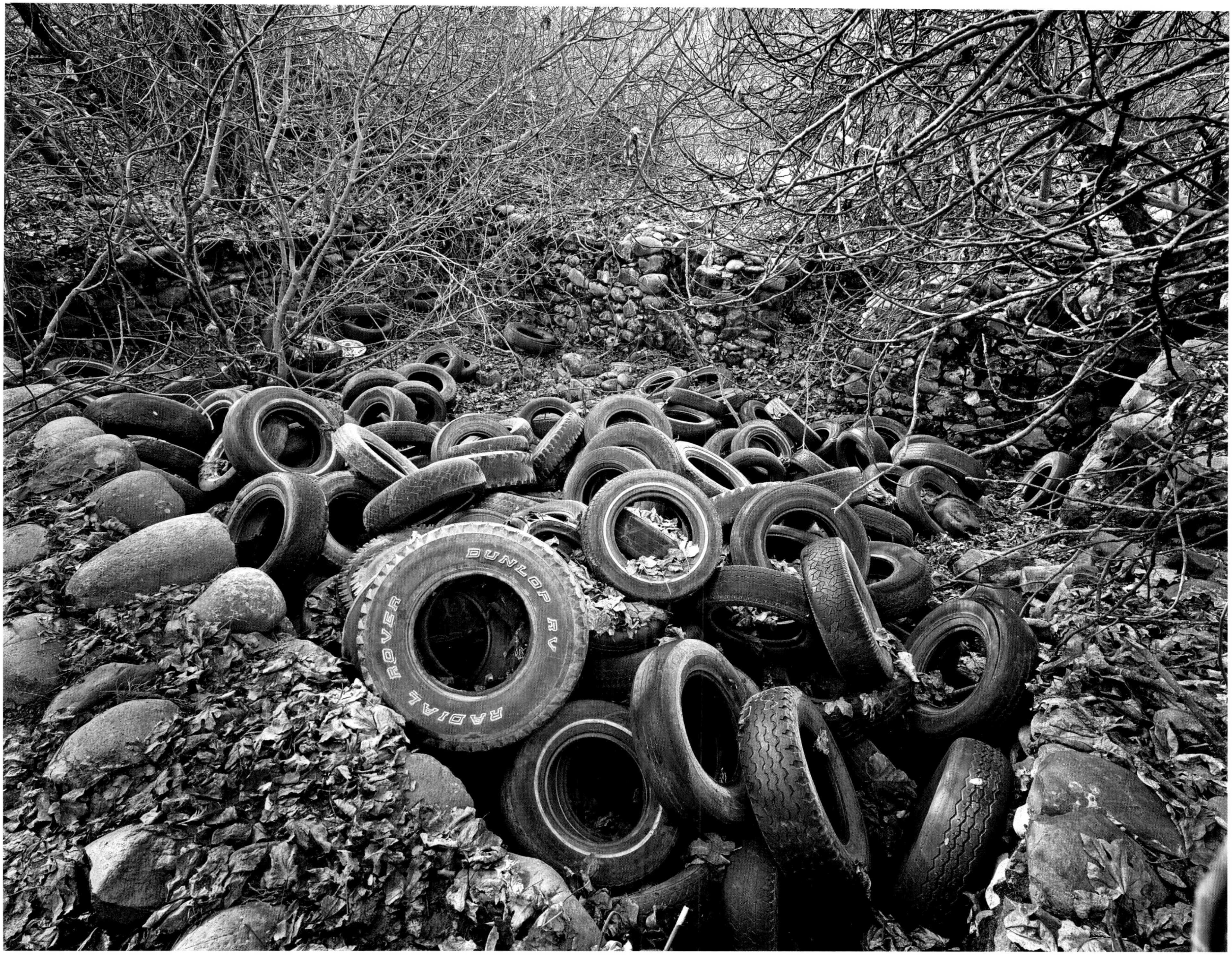

Foundation for a mining-camp whiskey still, now filled with tires, Butte Creek, 1990

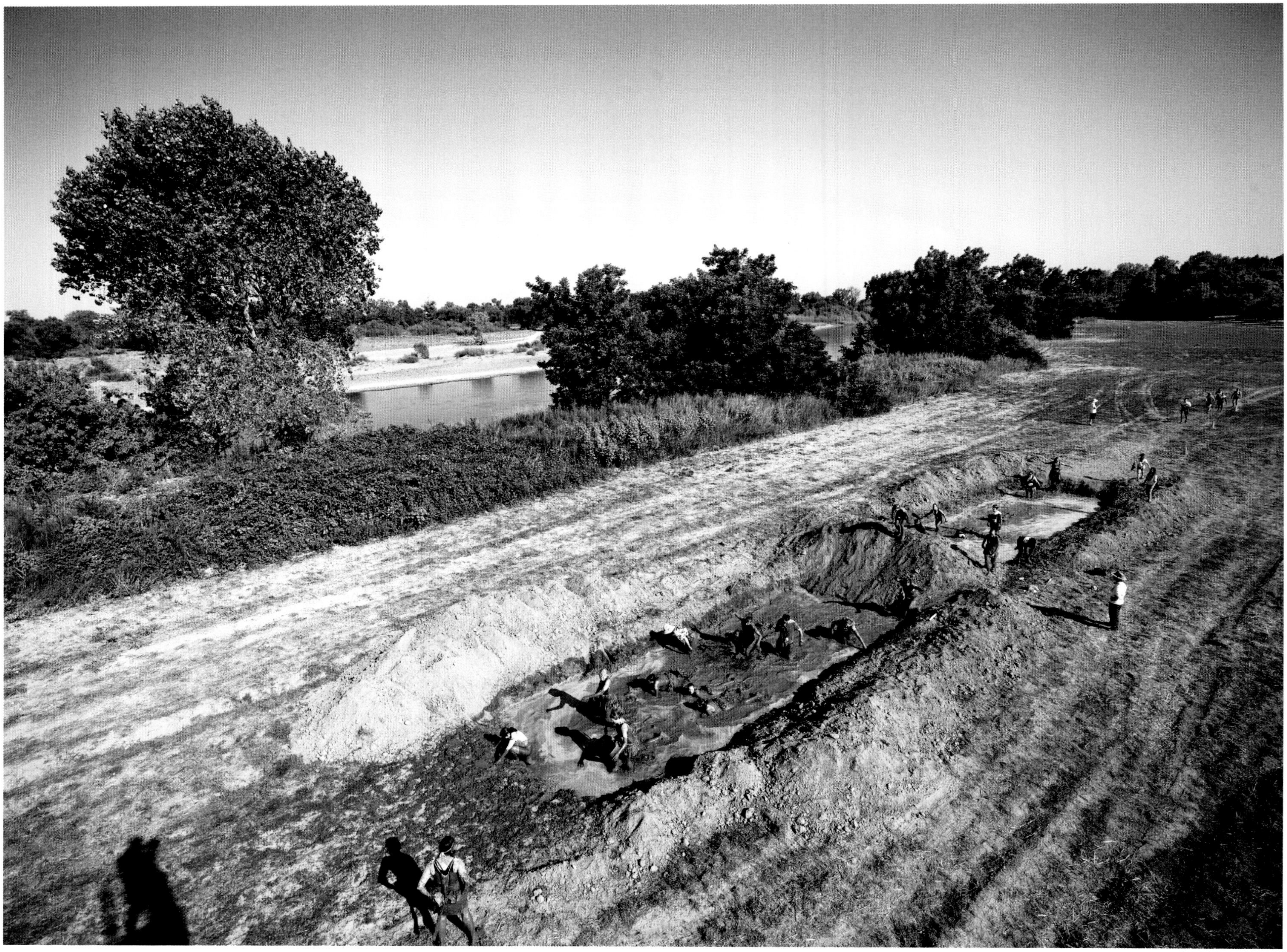

River Partners' Mudder Nature Challenge pits with Sacramento River, 2011

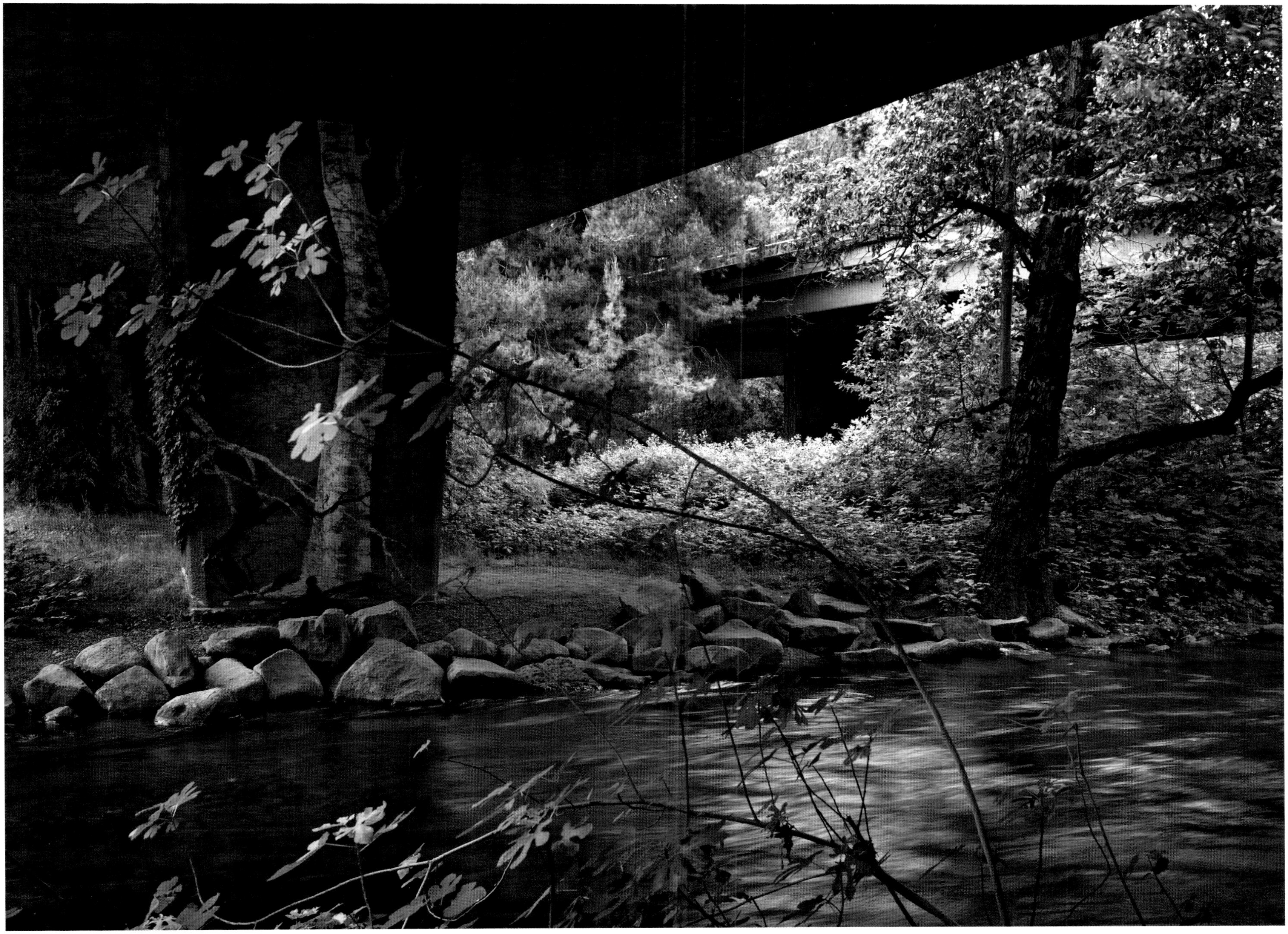

Greg Payne's painted sycamores on bridge pillars, Chico Creek, 2012

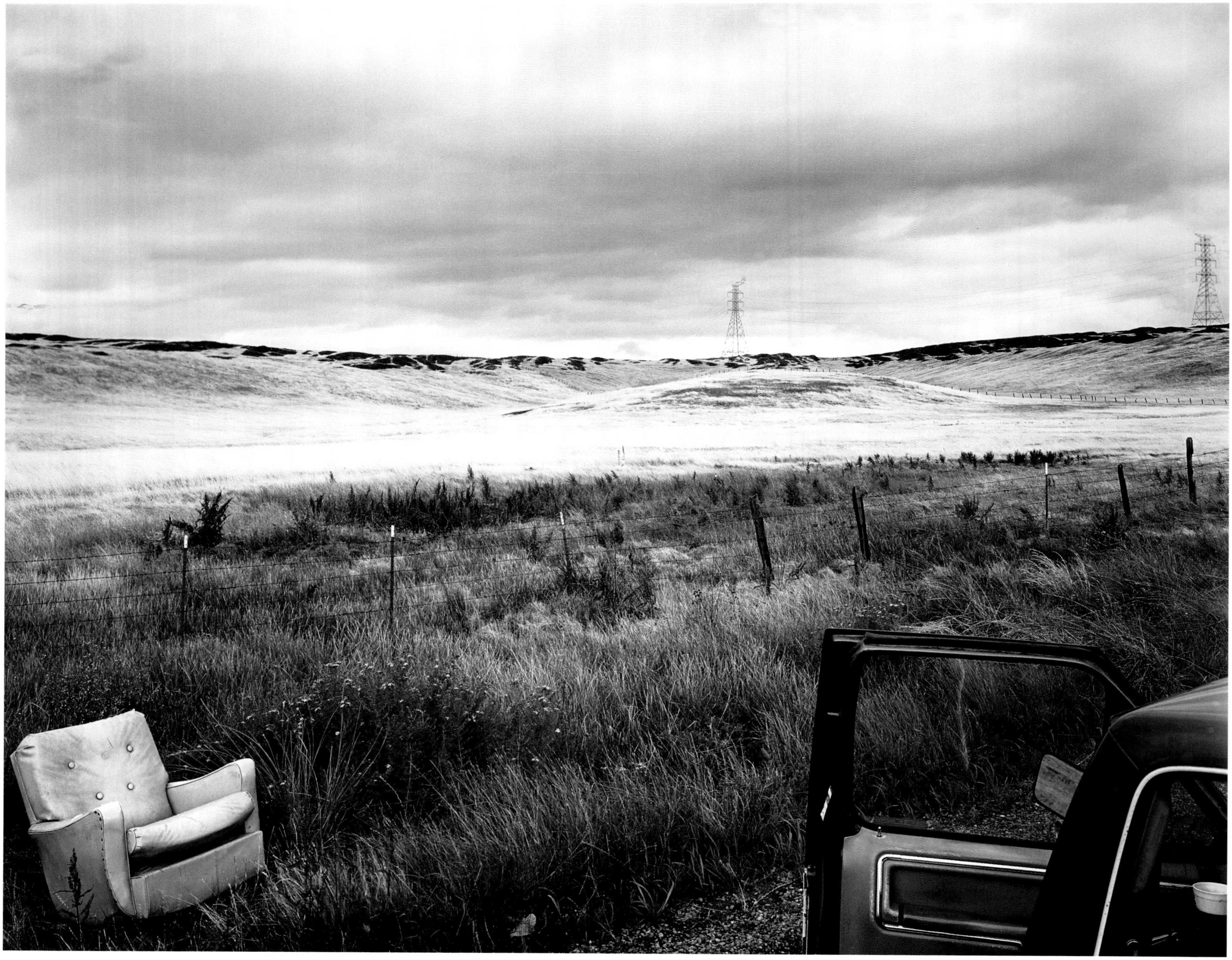

Rest stop (not photographable is the groundwater so highly contested in courtrooms), Butte County, 1993

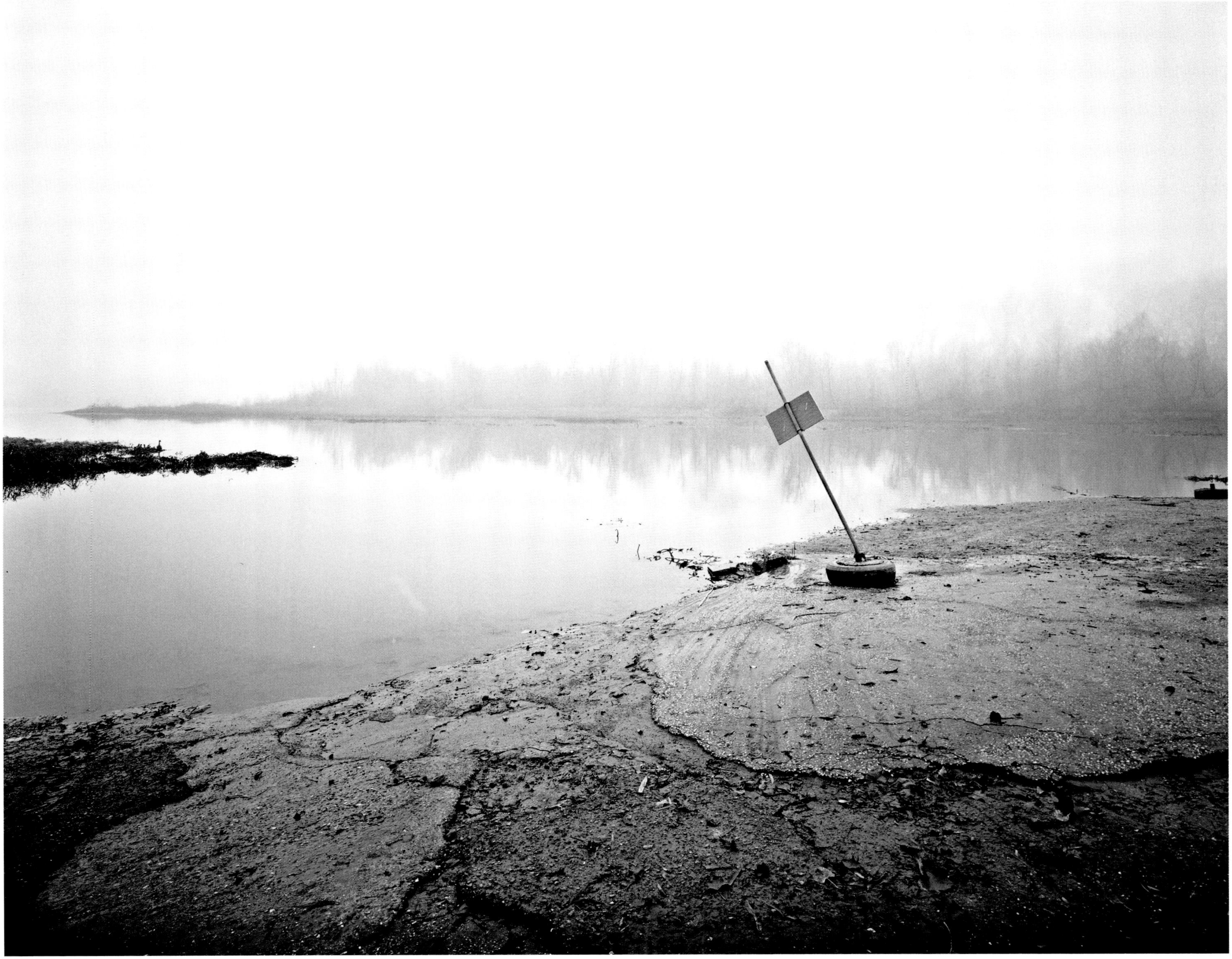

Sign, 2005

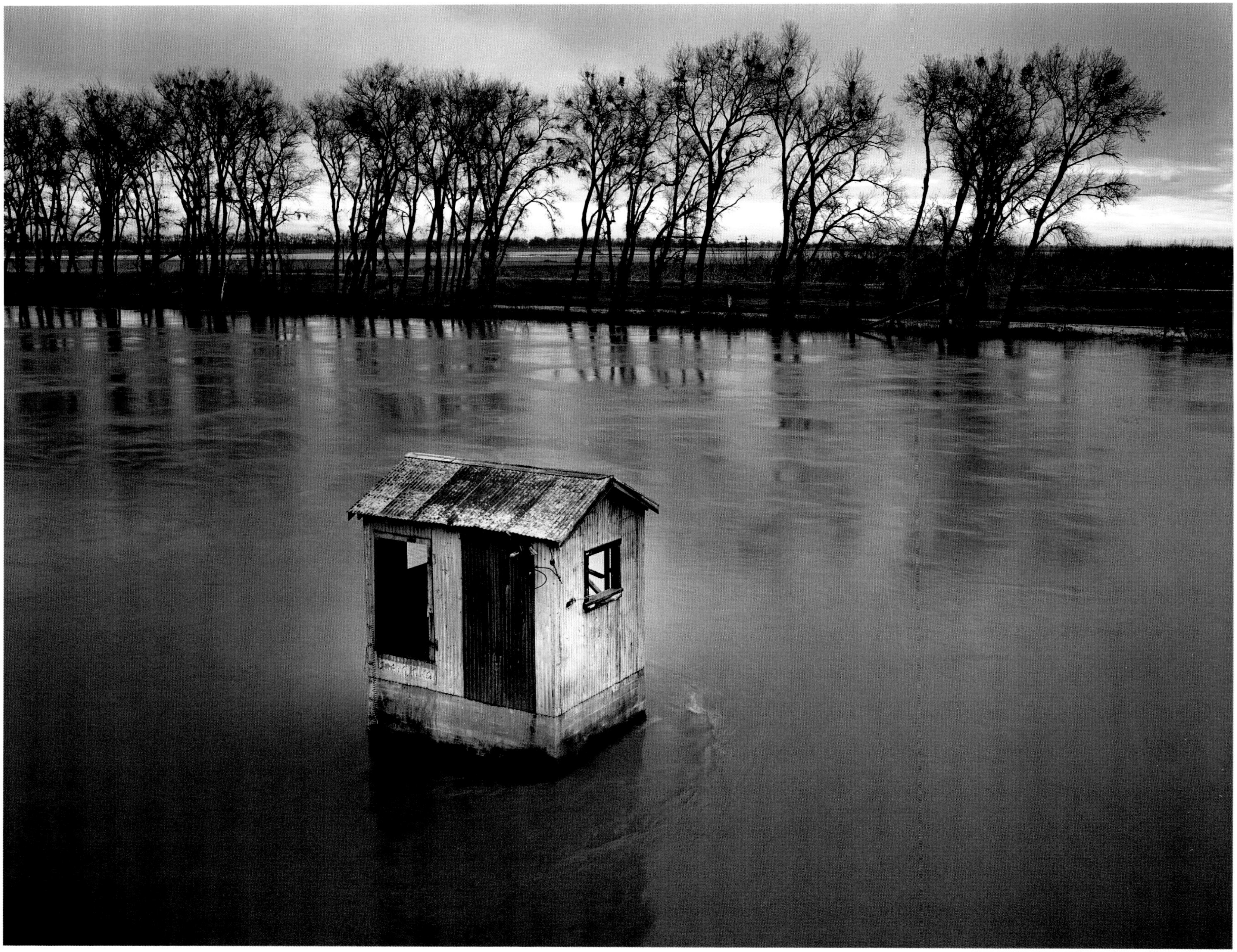

Submerged pump house with levee and surviving fossil tree stand in background, 1997

Surviving fossil tree stand at Scotty's Landing, 2005 »

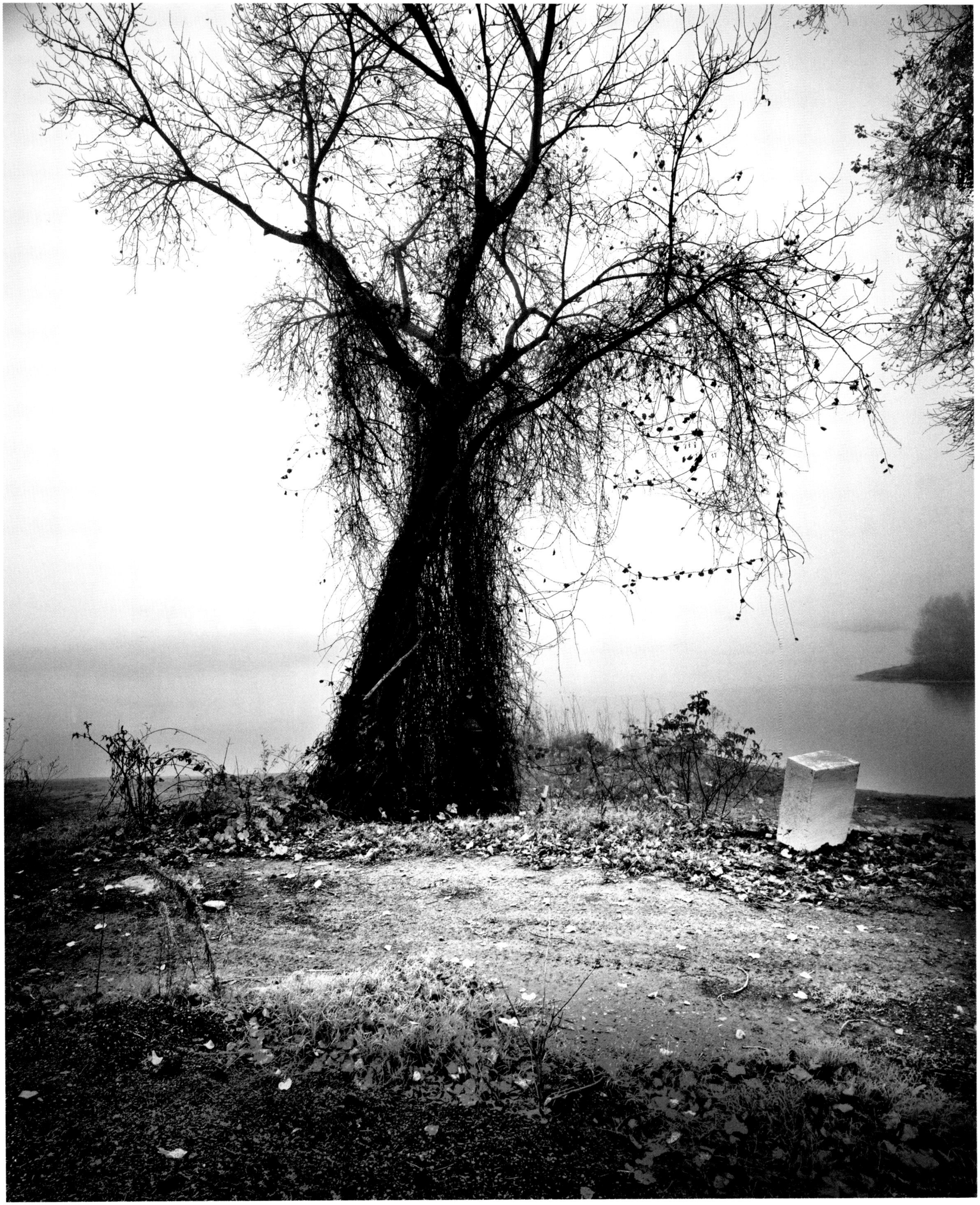

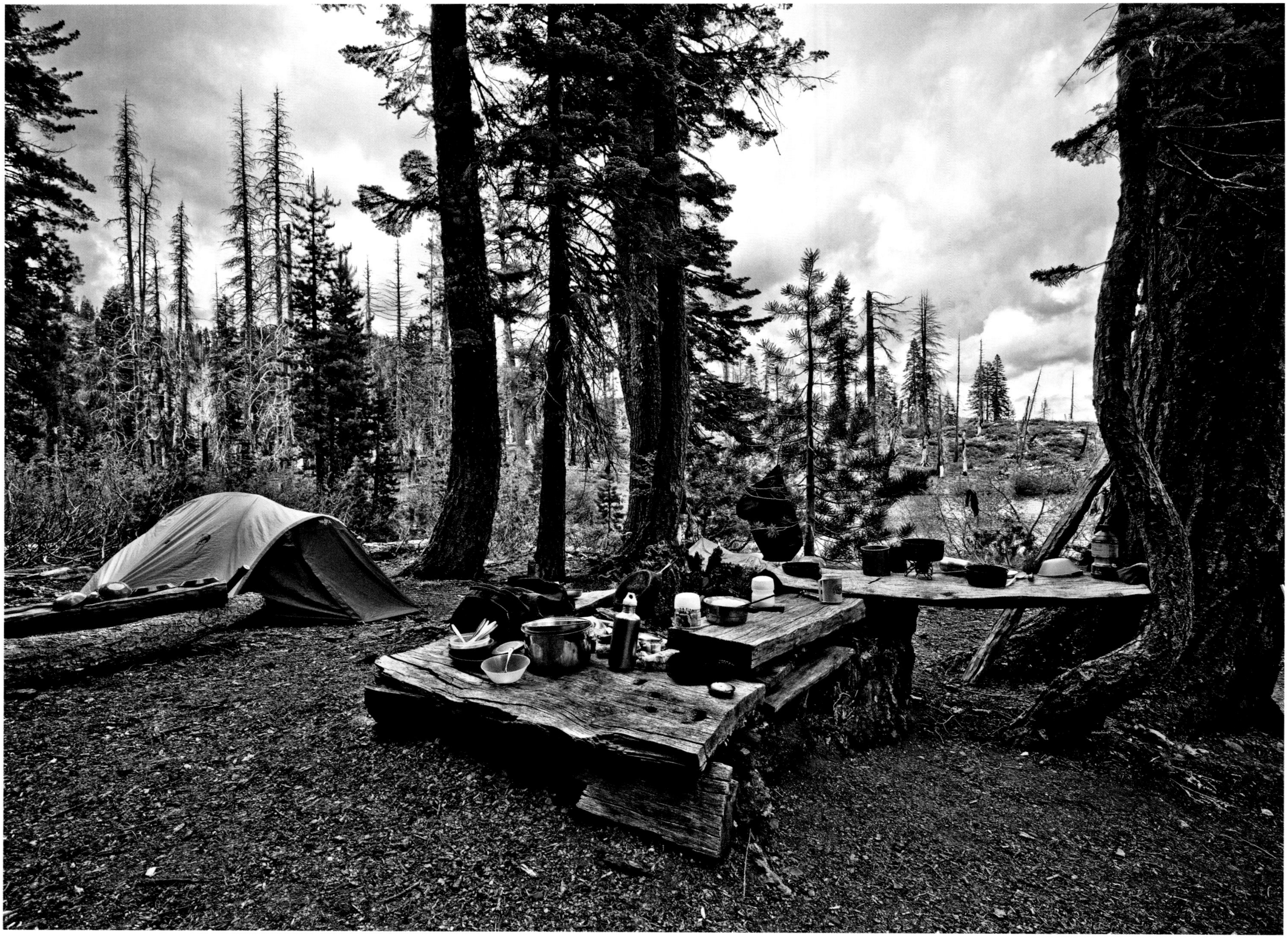

Camping in the Sierra, 2010

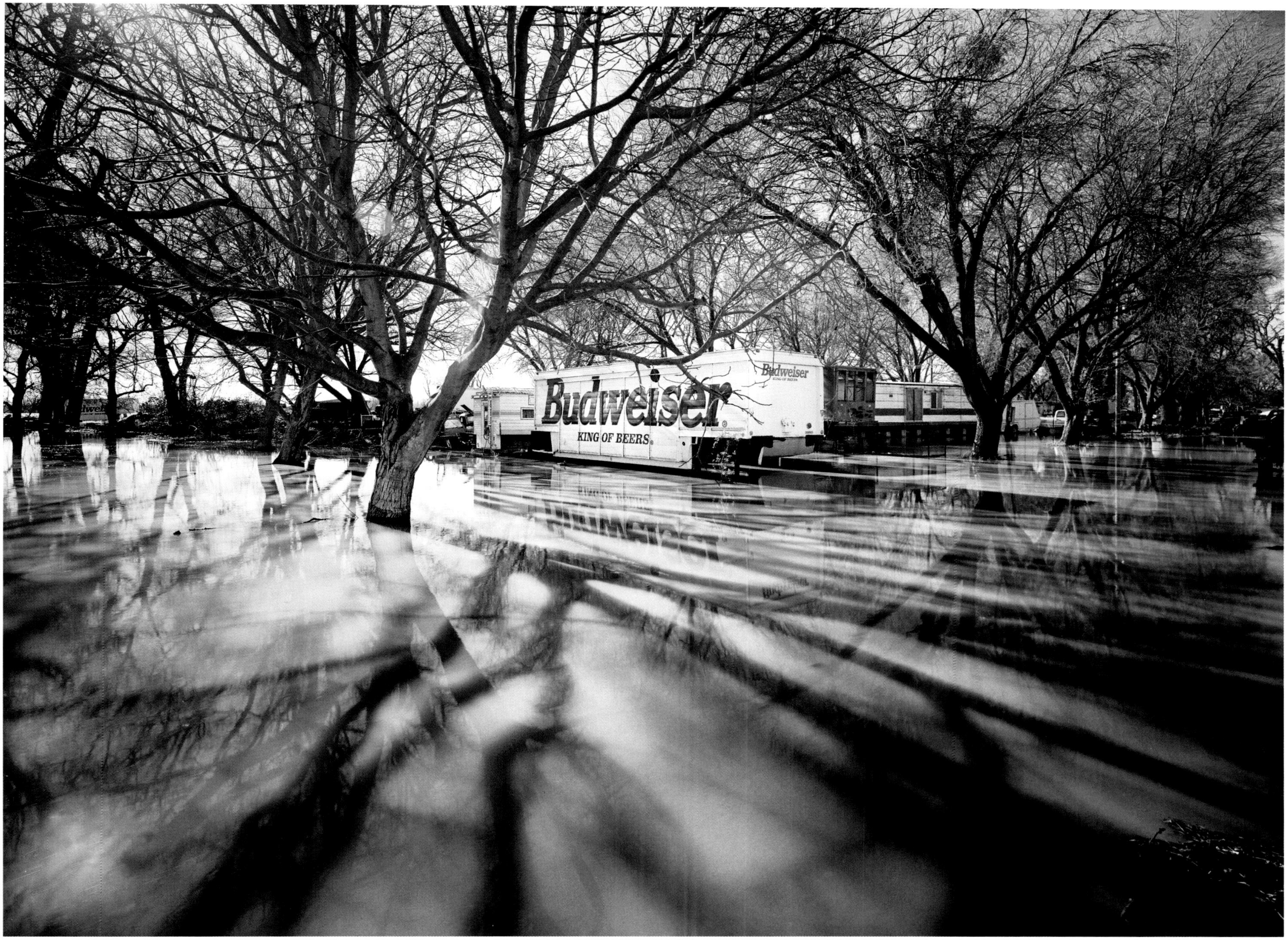

Budweiser truck during high water at Scotty's Landing, 2005

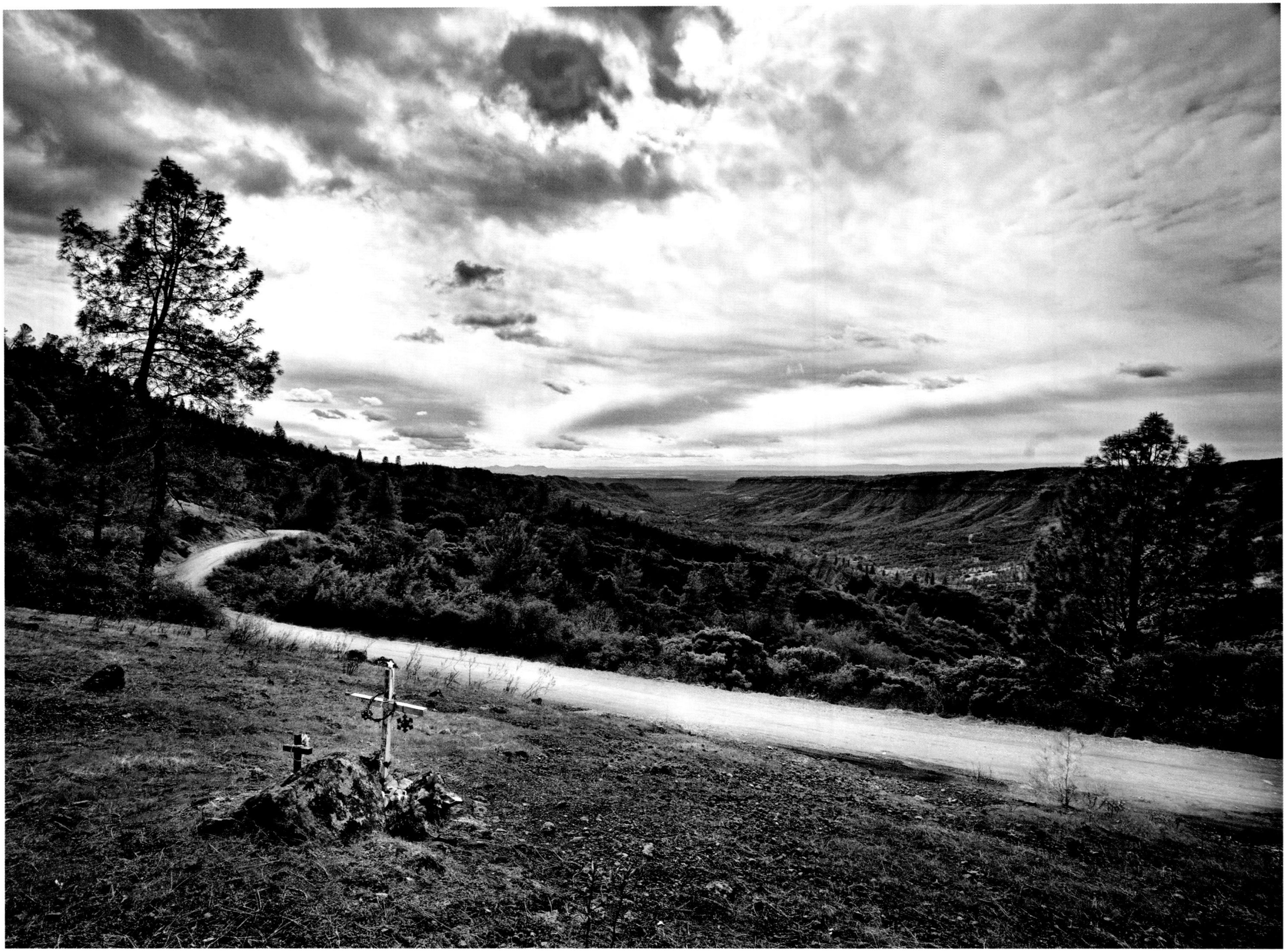

Butte Creek Canyon as viewed from Centerville Road to Magalia, 2012

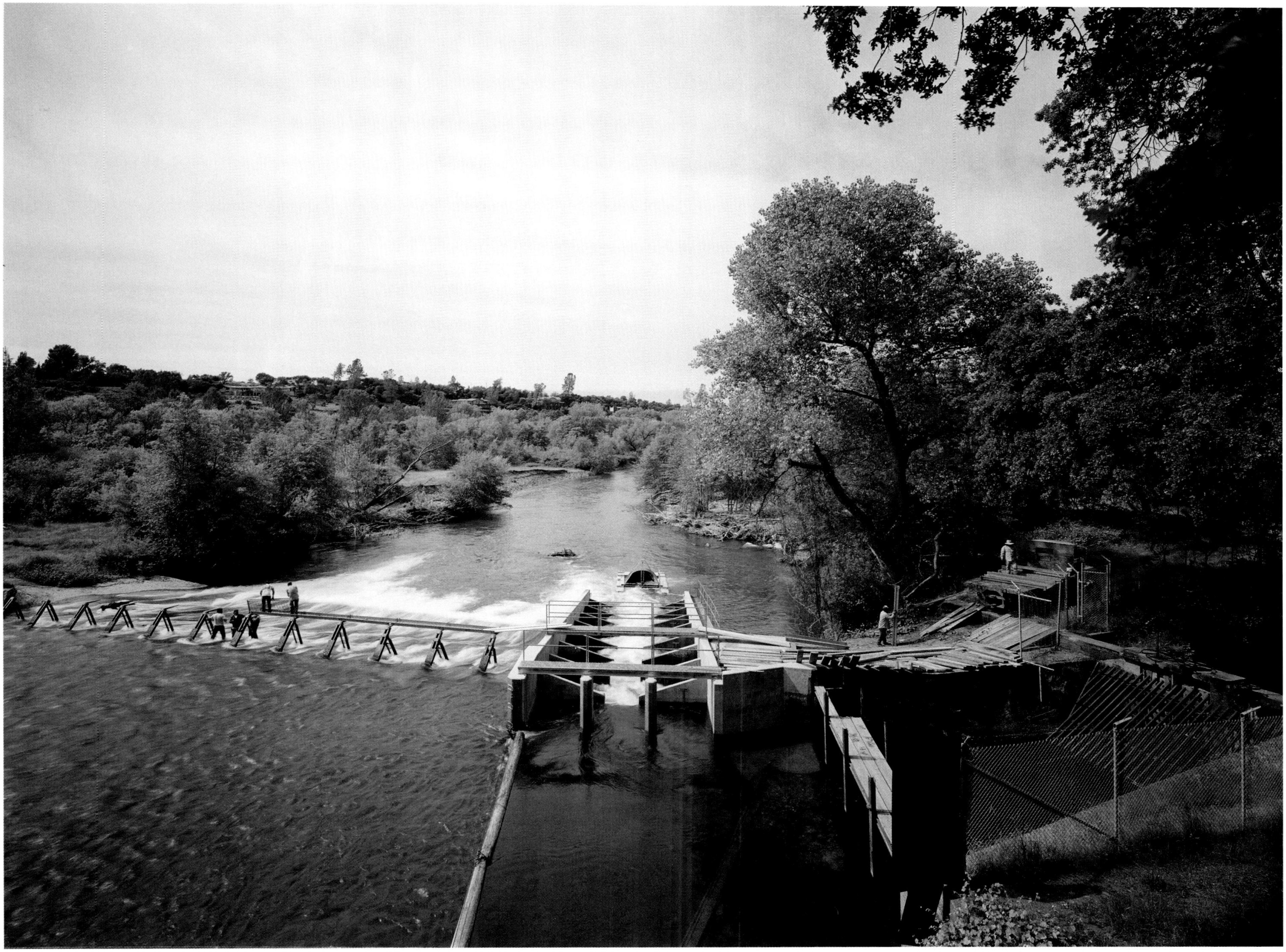

Workers pulling flashboards on the Parrott-Phelan Diversion Dam after the installation of a fish ladder, Butte Creek, 1996

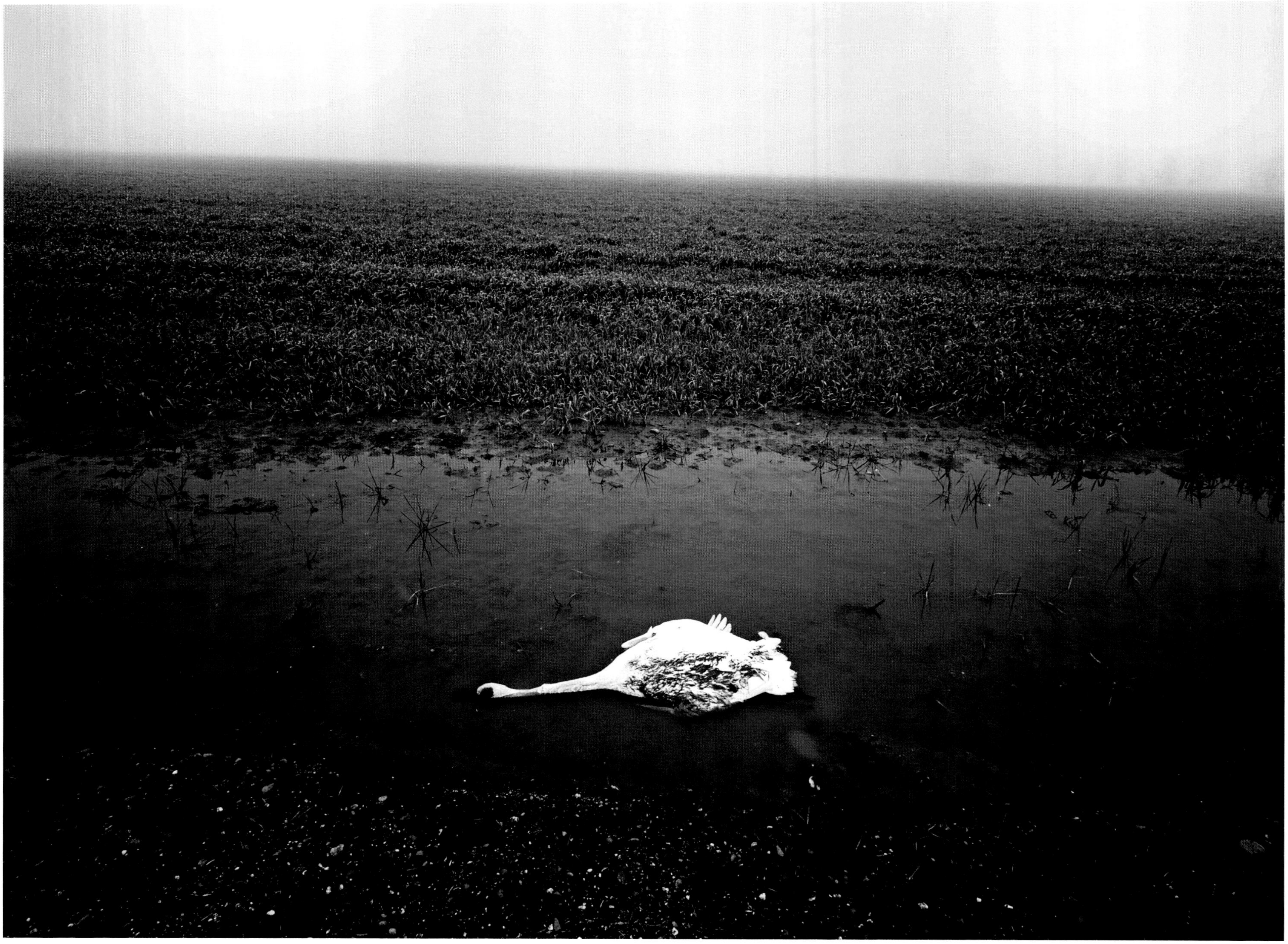

Trumpeter swan, 1994

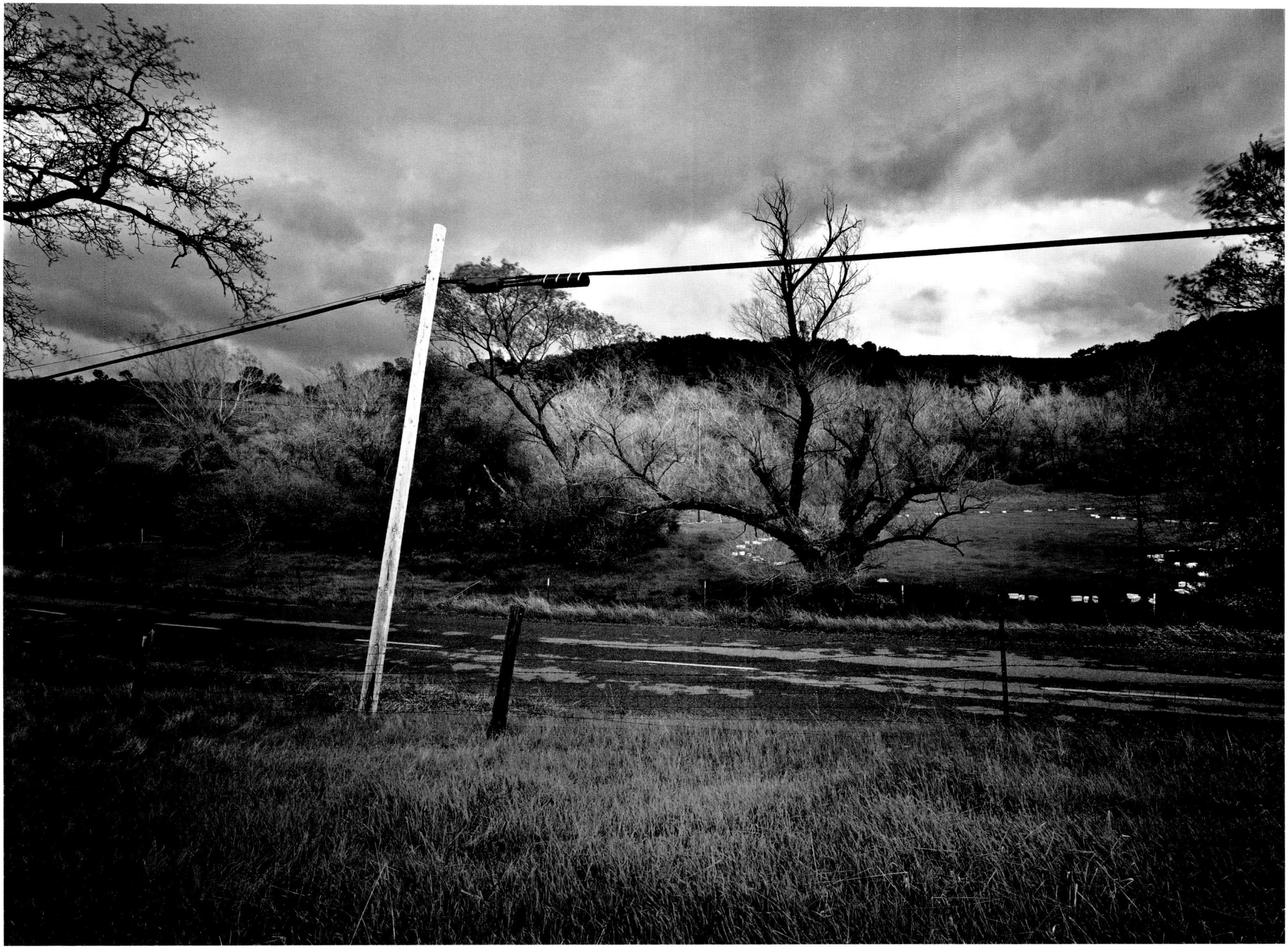

Nucleus beehives atop a leveled dredge site, Butte Creek, 1993

RESOURCES

Agricultural Issues Center. "Water Supply and Demand." White Papers on California Agricultural Issues. Davis, California: University of California, 2009. Accessed March 30, 2012. aic.ucdavis.edu/publications/whitepapers.

Bailey, Edgar H., ed. *Bulletin 190: Geology of Northern California.* San Francisco: California Division of Mines and Geology, 1966.

Bakker, Elna. *An Island Called California: An Ecological Introduction to Its Natural Communities.* Berkeley: University of California Press, 1972.

Bates, Robert L. and Julia A. Jackson, eds. *Dictionary of Geological Terms.* 3rd ed. Garden City, New York: Anchor Press/Doubleday, 1983.

Beers, Terry and Emily Elrod. *Califauna: A Literary Field Guide.* Berkeley: Heyday Books, 2007.

California Department of Water Resources. *California State Water Project Atlas.* Sacramento: Department of Water Resources, 1999.

California Department of Water Resources. *Central Valley Flood Management Program: Flood Control Systems Status Report.* Sacramento: Department of Water Resources, 2011.

Center for Irrigation Technology. *Agricultural Water Use in California: A 2011 Update.* Fresno, California: California State University, 2011.

Cunningham, Laura. *A State of Change: Forgotten Landscapes of California.* Berkeley: Heyday Books, 2010.

Dow, Kirstin and Thomas E. Downing. *The Atlas of Climate Change: Mapping the World's Greatest Challenge.* Berkeley: University of California Press, 2007.

Flores, Dan. "The Great Plains 'Wilderness' as a Human-Shaped Environment." *Great Plains Research: A Journal of Natural and Social Science* 9 (Fall 1999): 343–55.

Gleick, Peter, Penn Loh, Santos V. Gomez, and Jason Morrison. *California Water 2020: A Sustainable Vision.* Oakland: Pacific Institute for Studies in Development, Environment, and Security, 1995.

Kahrl, William. *The California Water Atlas.* Sacramento: Department of Water Resources, 1979.

Lawton, Rebecca, Diana Lawton, and Susan Panttaja. *Discover Nature in the Rocks: Things to Know and Things to Do.* Mechanicsburg, Pennsylvania: Stackpole Books, 1997.

Leopold, Luna B., M. Gordon Wolman, and John P. Miller. *Fluvial Processes in Geomorphology.* New York: Dover Publications, Inc., 1995.

Margolin, Malcolm, ed. *The Way We Lived: California Indian Reminiscences, Stories, and Songs.* Berkeley: Heyday Books, 1981.

McClurg, Sue. *Water and the Shaping of California.* Berkeley: Water Education Foundation and Heyday Books, 2000.

Micheli, Lisa, Lorraine Flint, Alan Flint, Morgan Kennedy, Stuart Weiss, and Ryan Branciforte. *Adapting to Climate Change: State of the Science for North Bay Watersheds (A Guide for Managers).* Santa Rosa, California: Pepperwood Preserve, 2010.

Northwest Regional Office National Oceanic and Atmospheric Administration. "Salmon Populations." Accessed May 27, 2012. http://www.nwr.noaa.gov/ESA-Salmon-Listings/Salmon-Populations.

Newton, Jim. "Water Ethics and a Peripheral Canal." *Los Angeles Times*, June 25, 2012. www.latimes.com.

Reisner, Marc. *Cadillac Desert: The American West and Its Disappearing Water.* New York: Viking Penguin Inc., 1986.

River Partners. "Sacramento River Projects." Accessed June 23, 2012. http://riverpartners.org/where-we-work/projects-sacramento.html.

"Sacramento River." Wikipedia: The Free Encyclopedia. Accessed May 27, 2012. en.wikipedia.org/wiki/Sacramento_River.

Sacramento River Preservation Trust. "History." Accessed June 23, 2012. http://www.sacrivertrust.org/aboutus/history.htm.

Snyder, Susan, ed. *Bear in Mind: The California Grizzly.* Berkeley: Heyday Books, 2003.

U.S. Department of the Interior, Bureau of Reclamation. "Central Valley Project Improvement Act." Reclamation: Mid-Pacific Region. Accessed April 20, 2012. www.usbr.gov/mp/cvpia.

U.S. Environmental Protection Agency and California Department of Water Resources. *Climate Change Handbook for Regional Water Planning.* Sacramento: Department of Water Resources, 2011.

U.S. Fish and Wildlife Service. "Refuge History." Sacramento River Wildlife Refuge Complex. Last modified August 17, 2009. www.fws.gov/sacramentovalleyrefuges/mgmt_history.html.

U.S. Geological Survey. "Sacramento River Basin National Water Quality Assessment Program." California Water Science Center. Last modified June 14, 2011. http://ca.water.usgs.gov/sac_nawqa/enviroset.html.

Wallace, David Rains. "Where the Elk and the Antelope Played: Reflections on Bay Area Megafauna." *Bay Nature,* January 1, 2004. baynature.org/articles/jan-mar-2004/where-the-elk-and-the-antelope-played.

Weisser, Matt. "Findings Offer New View of Early Indians' Life along Sacramento River." *The Sacramento Bee,* April 4, 2011. www.sacbee.com/2011/04/04/3525654/findings-offer-new-view-of-early.html.

Wolff, Jane. *Delta Primer: A Field Guide to the California Delta.* San Francisco: William Stout Publishers, 2003.

ABOUT THE AUTHORS

GEOFF FRICKER's photographs are housed in the permanent collections of a diverse set of museums across the United States, including the San Francisco Museum of Modern Art, the Oakland Museum of California, the Crocker Art Gallery, the Library of Congress, and other cultural institutions in California, Hawaii, Arizona, Nevada, and Texas. He has received numerous awards and grants to document watersheds. He holds an MFA in photography from the San Francisco Art Institute. Geoff lives with his wife, Sandee, on the banks of Butte Creek, a major salmon tributary of the Sacramento River.

REBECCA LAWTON was one of the first women river guides on the Colorado in the Grand Canyon as well other rivers in the West. Since 1974 she has teamed with geomorphologists, paleontologists, and geographers to study the movement of silt, clay, sand, fossils, and other sedimentary particles in streams. Her writing honors include the inaugural Ellen Meloy Fund Award for Desert Writers and residencies at Hedgebrook and The Island Institute. Rebecca's essay collection, *Reading Water: Lessons from the River,* was a *San Francisco Chronicle* Bay Area Bestseller. Her novel, *Junction, Utah,* about the impacts of domestic oil, foreign war, and cultural differences on wilderness rivers and small-town American life, was published in early 2013. Currently Rebecca serves on the board of directors for Friends of the River. She lives with her husband, Paul, on Pequeño Creek in Sonoma Valley.

HEYDAY
into California

ABOUT HEYDAY

Heyday is an independent, nonprofit publisher and unique cultural institution. We promote widespread awareness and celebration of California's many cultures, landscapes, and boundary-breaking ideas. Through our well-crafted books, public events, and innovative outreach programs we are building a vibrant community of readers, writers, and thinkers.

THANK YOU

It takes the collective effort of many to create a thriving literary culture. We are thankful to all the thoughtful people we have the privilege to engage with. Cheers to our writers, artists, editors, storytellers, designers, printers, bookstores, critics, cultural organizations, readers, and book lovers everywhere!

We are especially grateful for the generous funding we've received for our publications and programs during the past year from foundations and hundreds of individual donors. Major supporters include:

Anonymous (3); Alliance for California Traditional Arts; Arkay Foundation; Judy Avery; James J. Baechle; Paul Bancroft III; BayTree Fund; S. D. Bechtel, Jr. Foundation; Barbara Jean and Fred Berensmeier; Berkeley Civic Arts Program and Civic Arts Commission; Joan Berman; John Briscoe; Lewis and Sheana Butler; California Civil Liberties Public Education Program; Cal Humanities; California Indian Heritage Center Foundation; California State Parks Foundation; Keith Campbell Foundation; Candelaria Fund; John and Nancy Cassidy Family Foundation, through Silicon Valley Community Foundation; Charles Edwin Chase; Graham Chisholm; The Christensen Fund; Jon Christensen; Community Futures Collective; Compton Foundation; Creative Work Fund; Lawrence Crooks; Nik Dehejia; Frances Dinkelspiel and Gary Wayne; The Durfee Foundation; Earth Island Institute; The Fred Gellert Family Foundation; Fulfillco; The Wallace Alexander Gerbode Foundation; Nicola W. Gordon; Wanda Lee Graves and Stephen Duscha; David Guy; The Walter and Elise Haas Fund; Coke and James Hallowell; Stephen Hearst; Cindy Heitzman; Historic Resources Group; Sandra and Charles Hobson; Donna Ewald Huggins; Humboldt Area Foundation; JiJi Foundation; The James Irvine Foundation; Claudia Jurmain; Marty and Pamela Krasney; Guy Lampard and Suzanne Badenhoop; Christine Leefeldt, in celebration of Ernest Callenbach and Malcolm Margolin's friendship; Bernard and Josie Le Roy; Thomas Lockard; Thomas J. Long Foundation; Judith and Brad Lowry-Croul; Michael McCone; Nion McEvoy and Leslie Berriman; Giles W. and Elise G. Mead Foundation;

Michael Mitrani; Moore Family Foundation; Richard Nagler; National Endowment for the Arts; National Wildlife Federation; Native Cultures Fund; The Nature Conservancy; Nightingale Family Foundation; Northern California Water Association; The David and Lucile Packard Foundation; Panta Rhea Foundation; Alan Rosenus; The San Francisco Foundation; Greg Sarris; William Somerville; Martha Stanley; Roselyne Chroman Swig; Swinerton Family Fund; Sedge Thomson and Sylvia Brownrigg; TomKat Charitable Trust; The Roger J. and Madeleine Traynor Foundation; Lisa Van Cleef and Mark Gunson; Patricia Wakida; John Wiley & Sons, Inc.; Peter Booth Wiley and Valerie Barth; Dean Witter Foundation; The Work-in-Progress Fund of Tides Foundation; and Yocha Dehe Community Fund.

GETTING INVOLVED

To learn more about our publications, events, membership club, and other ways you can participate, please visit www.heydaybooks.com.